The

ALCHEMY & ALCHEMISTS

www.pocketessentials.com

First published in Great Britain 2001 by Pocket Essentials, 18 Coleswood Road, Harpenden, Herts, AL5 1EQ

Distributed in the USA by Trafalgar Square Publishing, PO Box 257, Howe Hill Road, North Pomfret, Vermont 05053

Series Editor: Nick Rennison

A CIP catalogue record for this book is available from the British Library.

ISBN 1-903047-52-8

9 8 7 6 5 4 3 2 1

Book typeset by Pdunk

Printed and bound by Cox & Wyman

For My Parents

Acknowledgements

Nick Rennison for his encouragement and support; Mike Paine, for going the distance at the same time; the Venerable Nik Harding, for lending me certain ancient books when I had need of them; the Wardens of the Church at Otterhampton, Somerset, who provided fascinating insights into the life and work of Thomas Charnock, and showed me the lane that he lived on; the Hather Family, for driving me out to Charnock's neck of the woods; and my sister Lois, for her love.

CONTENTS

For verily I say unto you, that whosoever shall say unto this mountain, 'Be thou removed, and be thou cast into the sea'; and shall not doubt in his heart, but shall believe that those things which he saith shall come to pass; he shall have whatsoever he saith. Therefore I say unto you, what things soever ye desire, when ye pray, believe that ye receive them, and ye shall have them.

Mark 11, 23-24

Medieval people weren't seeking logical explanations; they were seeking harmony.

Midas Dekkers, *The Way of All Flesh*

Be ye transmuted into living philosophical stones.

Eugenius Philalethes [Thomas Vaughan]

Preface

This book will attempt to give a brief history of some of the main themes and practitioners of alchemy, the so-called Royal Art. It is not an exhaustive or definitive work on the subject, and, necessarily limited by length, I have felt that the best approach to take was that of a personal, poetic response to what has been called 'the art of transformation'. Alchemy could also be described as 'the art of possibilities', and I feel that what it has of value lies in this area. The alchemists were not living in a world limited by materialism and empiricism, and realised that nothing, including ourselves, is immutable. They saw that change is a natural process, and with openness, imagination and perseverance, everything could be changed into something of greater worth. Alchemy stresses that the inner and the outer must be joined together, the body with the mind or spirit, and the individual with nature.

Almost everything in alchemy is open to debate, refutation and denial; some alchemists are revered by some as masters and regarded as charlatans by others. Where there are several 'explanations' for something (for example the wealth of Nicholas Flamel being due to a successful series of transmutations, or simple frugality and good business sense over many years) I have tended to favour the magical, alchemical version of events. I am prepared to give them the benefit of the doubt, whether they seemed to be genuine like Flamel, or rogues. The point that is important here is that once one accepts the *possibility* of transmutation, be it lead into gold, or an unhappy human being into a happy one, then, to paraphrase the words of one of the most mysterious of all alchemists, John Kellerman, we have the world in our hands.

Sean Martin
Feast Day of St John of Rila
2000

Introduction

The History of an Error, or the Mightiest Secret that a Man can Possess?

Alchemy has been with us since the beginning of recorded history. It has been present in almost every culture, from Old Kingdom Egypt and the China of Lao Tzu, through to the Greece of Alexander the Great and to the period of Islamic expansion, and from the islands of the Indonesian archipelago to the twilight world of Victorian occultism. It has been seen as a Divine Art, the highest gift of God, one that should only be practised by the sincere seeker and the pure of heart, and also as a fraudulent, delusional quest for wealth and worldly power. Probably more books have been written on alchemy than almost any other subject in human history. Ben Jonson famously satirized it in his play *The Alchemist* (1610), while the Brazilian writer Paolo Coelho's fable-like book of the same title, which seems to be much more sympathetic to alchemy than Jonson, was one of the best-selling novels of the 1990's.

What are we to make of this vast sea of material? We are all familiar with the supposed aims of alchemy, that of turning lead into gold, and we have all been brought up to believe that it is impossible. More recently than Ben Jonson, we have seen the foibles and failures of the science brilliantly sent up by Ben Elton and Richard Curtis in *Blackadder the Second*, where Percy, much to Edmund's withering disappointment, creates not a lump of gold, but a lump of green.

Alchemists have not surprisingly been seen as charlatans out to dupe wealthy potential backers, as Subtle and Face do so brilliantly in Jonson's play. Frequently, rulers have had to outlaw the art, as Pope John XXII did in his bull of 1317:

> Poor themselves, the alchemists promise riches which are not forthcoming; wise also in their own conceit, they fall into the ditch which they themselves have digged.

Pope John was not the only one. As early as 144 BC, the Chinese Emperor issued an edict forbidding the manufacture of gold. Similar decrees were issued periodically throughout history: China banned it again in 60 BC; the Roman Emperor Diocletian in AD 296; while Henry IV made it illegal in England in 1403.

Yet behind the litany of charlatans, or the hopelessly misguided, is a tradition that claims to give one access to the deepest mysteries of both nature and the self. For every king and emperor who outlawed the art, there are numerous examples of kings, popes and nobles who either practised alchemy or at the very least encouraged it. Sylvester II, the first French pope (999-1003), was alleged to have made a talking head through magical means and seven or eight years before his election to the pontificate had to swear before a council at Amiens that he wasn't a sorcerer. King James IV of Scotland (1473-1513) was supposed to have carried out experiments in Stirling Castle; Queen Christina of Sweden abdicated in order to study alchemy; Charles II of England had his own private laboratory built beneath the royal bed chamber, and, perhaps most famously of all, the Holy Roman Emperor Rudolph II (1576-1612) was so obsessed by the Art that he is supposed to have neglected affairs of state in his quest for transmutation.

And not only kings and princes lent their time and money to the study of alchemy: eminent mediaeval scholars such as Albertus Magnus, Moses Maimonides, Roger Bacon and Thomas Aquinas all shared an interest in the quest for transmutation. Perhaps more interestingly, so did most of the founders of modern science, men such as Jan Battista van Helmont, who discovered gases and was one of the pioneers of modern chemistry, Robert Boyle and Sir Isaac Newton.

Were all these great minds simply mistaken in their studies of alchemy? Perhaps not all of them saw it as a quest to produce gold. For some, the gold was already there, in the alchemist's heart, and the laboratory work was the ritual, meditative element that nourished this inner wealth. As modern science began to emerge in the late seventeenth century, so alchemy turned increasingly away from the laboratory aspect and became an increasingly mystical practice. To Fulcanelli, the most enigmatic alchemist of the twentieth century, alchemy was 'the art of transformation.'

For the Swiss psychologist Carl Jung (1875-1961) and his followers, the alchemist was not so much trying to create precious metals in the laboratory, as to redeem matter. Jung, who has done so much to redeem the entire subject, wrote:

> The alchemical operations were real, only this reality was not physical but psychological. Alchemy represents the projection of a drama both cosmic and spiritual in laboratory terms. The opus magnum had two aims: the rescue of the human soul and the salvation of the cosmos.

Between the two poles of the fraud seeking gold on the one hand and a purely psychological approach on the other, alchemy has long been seen as a precursor to modern chemistry, which is perhaps the kindest description of the Art that orthodox history will afford. Discoveries were made, perhaps by accident, amidst the noxious fumes of the laboratory. Some no doubt were trying to make gold, or perfect metals, as they might have phrased it. Others, possibly those artists of a more inward-looking nature, would be attempting, as Jung put it, to rescue the soul and the cosmos.

Roger Bacon discovered the properties of antimony, while the Arabs invented the camera obscura. Alcoholic distillation, phosphorous, porcelain and sodium nitrate were all the products of the alchemist's lab. Paracelsus spoke of the circulation of the blood one hundred years before it was 'officially' recognised, while Michael Sendivogius discovered oxygen over a century and a half before Joseph Priestly's discovery in 1774. The French alchemist Tiphaigne de la Roche understood the process of fixing images, and may well have been taking photographs as early as the 1720s, a century ahead of Dageurre and Fox Talbot. Sir Isaac Newton, a lifelong practicing alchemist, spoke of the Art as concealing secrets that would be dangerous should they fall into the wrong hands, which has led some to believe that he understood, or intuited, the secrets of nuclear power. (In 1937 there was a famous meeting between the French writer Jacques Bergier and a mysterious stranger, who was introduced to him as an alchemist, and whom he took to be Fulcanelli; the stranger told Bergier of the dangers of atomic research before disappearing. Bergier never saw him again.)

Since Jung alchemy has been taken much more seriously in academic circles, and it is now clear that the traditional view that alchemy was merely a form of early chemistry can no longer be held for the subject as a whole. Alchemists can safely be said to have invented the laboratory, and all laboratory implements, which modern chemistry did inherit, but there is more to the art than this.

Alchemy can also be seen as a spiritual path, undertaken by students studying under an adept or master. From its beginnings in China and Egypt, this is how alchemy was studied. With the beginnings of modern science in the seventeenth century, alchemy became more and more an inner discipline. Writers such as Jakob Boehme (1575-1624) and Thomas Vaughan (1621-1665) were almost certainly total strangers to the lab. These writers often equated the Philosopher's Stone with Christ, and can be seen as mystics as much as alchemists. (It is no surprise to learn that Thomas Vaughan was the twin brother of the great metaphysical poet Henry Vaughan, who also employed alchemical imagery in his works on occasion.)

The mystical branch of alchemy produced countless allegories of spiritual development. Pre-dating John Bunyan by fifty years or so, is the remarkable *The Labyrinth of the World and the Paradise of the Heart* by J.A. Comenius. In this, the narrator is lead through an endless city where he witnesses all forms of human folly. In the end, he is saved only by his faith in God. In addition to Comenius, there are countless shorter works, such as Thomas Vaughan's *House of Light* and Tommasso de Campanella's *City of the Sun.*

Alchemical books were often profusely illustrated. Anyone familiar with the history of Western art will, however, find the images strangely timeless and possibly a little shocking. There are pictures of people tearing their own hearts out, or of bodies being dismembered. People are shown as cripples, as if to mirror their crippled or limited understanding of the Art, or the world (if indeed there is ultimately a difference.) Sexual imagery is strong, with the alchemical archetypes of the King and Queen frequently portrayed engaging in sexual intercourse. In others, a man has a tree growing out of his body where his penis should be. Men and women merge, not just in the sexual act, but to become hermaphrodites.

In seems clear that these images were to be read or meditated upon, not merely looked at. Jung believed that, whatever sort of gold the alchemists were looking for, they had accidentally discovered the unconscious, and that their frequently strong, challenging images were, in fact, portraits of various states of consciousness that could lead us into a greater understanding of ourselves.

There is also a medical side to the Art. The great Swiss physician and reformer Paracelsus (c.1493-1541) worked largely in medicine, and in his alchemy can be seen the origins of modern homeopathy. This tradition claims its modern disciples in Archibald Cockren and Armand Barbault, both of whom saw it as a valuable complementary medicine. The popularity of alternative medicines is a testament to how powerful the ideas of Paracelsus have been. Anyone who has taken a homeopathic remedy has dipped a toe into the waters of alchemy without realising it.

Alchemy can also be seen to have played a part in the arts and culture. Jonson's play is no doubt the most celebrated example, and *Black-*

adder the Second the most recent, but alchemical ideas have also found their way into the works of Thomas Mann, the art of Joseph Beuys and Marcel Duchamp, countless books in the self-help sections of bookshops, and into Lindsay Clarke's Whitbread Prize-winning novel *The Chymical Wedding.* Dire Straits named an album after it, and the Art itself played a big part in the invention of opera, purposely designed as the 'ultimate art form'. The earliest operas (Peri's *Eurydice*, dating from 1600, is the earliest opera to survive complete) all dealt with themes that were essentially alchemical, and opera's first great figure, Monteverdi, was a practicing alchemist. It could be argued that the cinema, being an extension of the principles of opera, is also essentially alchemical.

It is perhaps all too easy to forget that the worldview that dominates the West, and has dismissed alchemy for so long, is a comparatively recent development. Newton saw the world much as his predecessors did, and has been described as the last Renaissance Man, 'the last of the Babylonian and Sumerian magi', as Maynard Keynes put it, because he saw Nature as a unity, a vast puzzle to be solved by the devout seeker. It is ironic that the world that Newton helped create has anything but a unified view of nature, and that its short-sightedness and materialistic greed threaten nature herself.

A further irony is that modern physics seems to be suggesting that such a worldview has had its day. Many of the theories of quantum physics will seemingly remain just that, as the money and energy needed to perform the experiments necessary to confirm the theses are simply beyond reach. Physics is therefore drifting evermore into a strange conceptual world not too far removed from that of alchemy. Alchemy would hold that everything in nature, and within the alchemist themselves, is related, and scientists are increasingly finding that there do seem to be connections between apparently isolated phenomena. Chaos theory's famous 'butterfly effect', where a butterfly flapping its wings in Australia can cause tidal waves in the Atlantic, is a case in point. In alchemy the idea of the unified worldview plays a central part: to the alchemist, even to the frauds or 'puffers', every part of the work was important. Not only was great care taken in all aspects of laboratory work, but attention was paid to the stars and phases of the moon. Dreams were recorded, intuition listened to. To the alchemist, there was nothing that was unrelated or irrelevant. There was no such thing as coincidence. Everything was part of the work.

This holistic view of the world has not been lost, and is still practiced by traditional societies the world over. Only in the West have we become cut off from this way of perceiving reality. This is perhaps why alchemy is still relevant to us. It deals with a power and a secret that each of us possess, that we need to rediscover and reclaim. We had it as children and we can have it again through work, study and inspiration.

Furthermore, quantum physics has shown that the role of the observer in an experiment is vital in determining its outcome; again, alchemy would seem to have come to these conclusions centuries earlier. And physics again has rediscovered ancient secrets in suggesting that the experimenter themselves can *actually manipulate matter* through strong visual imagination, which, to the alchemist, would be called Faith. (In a further irony, particle physics has shown that it is theoretically possible to turn lead into gold, if the number of molecules in the lead atom were to be increased. This would require vast expense and particle accelerators, which obviously the mediaeval alchemist of popular imagination did not have.) It seems that each individual, after all, may have a role to play in shaping the future and, indeed, of ensuring it happens at all.

1: Basic Ideas and Themes

Our view of alchemy is in the West is dominated by that of a medieval pre-chemist in a fume filled laboratory, a solitary figure working endlessly amidst bubbling flasks and studying ancient cryptic texts in a futile quest to turn lead into gold. They have perhaps received financial backing from a rich man who is keen to expand his wealth even further, but are really just wasting his money. Once it became known that alchemy supposedly had the power to make the practitioner immensely wealthy, the art began to attract frauds like flies to a dung heap. These were known as 'puffers', a reference to the bellows used to maintain the laboratory fire. No doubt they were full of hot air, too.

If the alchemist is genuine, he may inadvertently make a discovery that proves useful, but not in the way the alchemist or his backer had hoped. Alcoholic distillation was discovered in the alchemist's laboratory, for instance, and so was phosphorous. If the alchemist is not genuine, he will exhaust his backer's funds, and continue to move from town to town, impressing the rich and gullible with his laboratory demonstrations, and hoping that the authorities don't catch up with him.

The swindlers are the ones who usually find their way into literature and painting. Ben Jonson's play is perhaps the most famous example, overshadowing Chaucer's earlier *Canon's Yeoman's Tale*. Both works deal with false alchemy, and both Jonson and Chaucer seem to have had first hand knowledge of the subject; that they might both have lost money to fraudulent alchemists has been suggested as the motive behind the writing of both works.

The real alchemists, the workers unconcerned with becoming fabulously wealthy, are more shadowy figures. We could assume that they would be educated, given to keeping their own counsel, and hardworking. Their laboratories might have looked the same (at least to the uninitiated), but there would have been no sudden flashes of smoke or bright lights to impress would-be investors. Here work would have proceeded quietly and cautiously, almost like a Trappist monastery, with the alchemist swearing the lab assistants to strict secrecy. Equipment would have to be specially made without arousing the suspicions of the glassblower or potter. If word got out that one was involved in the alchemical work, there would be no telling what misfortunes might arise. If wealth was generated here, it was not the sort you could flaunt in front of others, or use to accumulate material goods for the sake of impressing your neighbours. The work was said to use only the simplest of materials, and was enigmatically likened to 'women's work and childrens' games.'

Much of the work would have consisted in observing changes in the flask as it was slowly heated. As this was traditionally one of the most secret, revered, and feared of human quests, we have little to compare it to directly. We could liken it to an artist in his studio, preparing the paints himself, and then using them to create art, something luminous and beautiful that would inspire others, and, in outliving the artist, partake of a form of immortality. We could also liken the alchemist to the doctor or apothecary, studying nature, selecting plants and herbs in order to prepare them in order to heal. The smith, frequently seen as a likely ancestor of the alchemist, occupied a similar position. In forging metals he was feared and revered as the solitary artist who stood on the doorway of worlds. Or the monk, studying texts, carefully illuminating them with either pictures or marginalia, providing a gloss that hinted of hidden treasure within something that one may read and otherwise miss. We may also see an equivalent in the quantum physicist, who has moved so far beyond experimental verification in investigating the properties of matter that he is working in areas formerly visited by the metaphysician and mystic.

Albertus Magnus (1193-1280) listed the qualities that the genuine alchemist must possess:

> First: He should be discreet and silent, revealing to no one the result of his operations.
>
> Second: He should reside in an isolated house in an isolated position.
>
> Third: He should choose his days and hours for labour with discretion.
>
> Fourth: He should have patience, diligence, and perseverance.
>
> Fifth: He should perform according to fixed rules.
>
> Sixth: He should only use vessels of glass or glazed earthenware.
>
> Seventh: He should be sufficiently rich to bear the expenses of his art.
>
> Eighth: He should avoid having anything to do with princes and noblemen.

As we shall see, would-be alchemists ignored this eighth precept at their peril.

The Goal of Alchemy

The goal of alchemy, in the West at least, was the production of the Philosopher's Stone, which would enable the alchemist to turn base metals such as lead into silver and gold. It was also said to bestow wealth and prolong life. It has been described as the Royal Art, or the Divine Art, or sometimes simply as the Art. Gold has traditionally been identified with divine powers because of its impermeability; it is resistant to both fire and water. Since the earliest times, gold and other, lesser, shiny metals such as silver, were regarded as having a divine origin and possessed the ability to overcome death: therefore, to own gold, or to try and emulate its properties, was to partake of divinity and immortality. It was also the colour of the sun, and silver that of the moon, the two great celestial powers that nurtured life by day and dreams - the life of the soul - by night.

Alchemists saw gold as growing in the belly of Mother Earth, and the aim of the art was to speed up natural processes in the laboratory. The fact that they saw gold and other metals as growing like plants over a very long period of time suggests that they intuitively understood the concept of geological time. They saw that there was a place within nature for alchemists (humanity) as the ones who should work with nature in order to perfect it and, because they saw within themselves the whole of creation, they were also perfecting themselves, and realising their true, divine natures.

During the Middle Ages, the Philosopher's Stone came to be equated with Christ. But there is another Christian parallel: in his Arthurian epic *Parzival*, Wolfram von Eschenbach describes the Holy Grail not as a cup or chalice, as is traditional, but as a stone beyond price, 'which the builders rejected, but which has become the corner stone.'

Laboratory Procedures

The work of the alchemist was known as the Great Work, and was sometimes divided into two stages, the Lesser Work, and the Greater Work. There is some dispute as to this, as each alchemist tended to see the process in their own way, and more importantly, to experience and live the work in their own way. The fourteenth century adept Basil Valentine spoke of the Twelve Keys of the work, while the fifteenth century English alchemist Sir George Ripley spoke of Twelve Gates. It was, moreover, a solitary work, with little chance for alchemists to meet and swap ideas for fear of capture by the authorities or ridicule.

Alchemy has always been an oral tradition, with the secret being passed on from master to pupil. Thomas Norton (fl. 1470s), it is said, rode for a hundred days in order to find a master who would teach him the secret. This would go some way to explain why no two alchemical manuscripts ever agreed on the precise nature and order of the work. There were as many ways to make the Stone as there were alchemists. Or, at least, as many variations on the process as there were alchemists: most, if not all, practitioners adhering to the basic black-white-red sequence prescribed by the legendary Egyptian adept Maria Prophetissa. This further stresses that alchemy was a solitary process: one that both took place away from the eyes of the curious, in the secret enclave of the laboratory, and also in the solitude of the alchemist's own being. Genuine alchemy, as with all other spiritual paths, needs human transmission in order for its magic to be fully grasped.

The threat of ridicule, harassment, kidnapping, torture and violent death was also no doubt an incentive to keep the work secret. An anonymous informant's information led to one fifteenth century English alchemist, Thomas Daulton, who was a monk at an abbey in Gloucestershire, being forced to the court of Edward IV in order to make gold for the king. Daulton was either unable to produce the elixir, or refused, and was kidnapped again, this time by one of Edward's nobles, Lord Herbert, who imprisoned Daulton for four years in Berkeley Castle. Daulton still refused to impart the secret, and was eventually released. He died a short time later, as did Herbert, who was killed at the battle of Tewkesbury.

The Great Work was often seen as being performed in one of two ways, the Dry Way and the Wet Way; some even spoke of further methods, called the Mixed Way and the Brief Way. These various methods refer to the supposed length of time of the work, and the temperatures required to heat the matter in the flask.

The Wet Way is described as the noblest of the ways, and is usually said to require the most time to accomplish. In the wet way, the temperatures needed to heat the retort are low: often much of the work is said to be carried out at body temperature, or at the temperature of a hen sitting on her eggs.

The Dry Way utilises an oven, and much greater temperatures, sometimes of up to 1000° centigrade. This is traditionally said to be a difficult method, and one that should only be attempted in conjunction with a master. This is the method said to have been used by Fulcanelli.

The Mixed Way, perhaps not surprisingly, incorporates procedures from both the Wet and Dry Ways, and was supposedly the method used by Albertus Magnus, Nicholas Flamel and Eirenaeus Philalethes.

The Brief Way is, as its name suggests, a direct, short path, a sort of alchemical dzogchen, that is said to take as little as four days to accomplish.

In Jungian terms, the length of the process would depend upon how many things the individual has to work through in order to achieve a sense of wholeness and balance.

Alchemy, at least in the traditional sense, is a very practical work that one cannot learn from books (we shall later see how a purely philosophical branch evolved that had little to do with laboratory work).

The Lesser Work: Nigredo

The alchemical work began with the first matter, or first agent, being placed into the retort at an astrologically suitable time (usually the spring, under the signs of Aries and Taurus). Astrology has always been important in alchemy, as it is one of the main tools by which an alchemist can try and work in harmony with natural rhythms and cycles.

What was it that the alchemists put into their baths and stills? No one knows. The word for alchemy, *al-kimia*, is, according to some authorities, taken to mean 'the art of the black land', and we can hazard a guess that, at least for the Egyptian alchemists, the first matter or prima materia may have been the black soil that was fed by the Nile and its tributaries; Egypt was called the black land precisely because of its well-nourished Nile soil. Later alchemists would write that 'the matter of our work is everywhere present', and also that it was 'worthless' and 'vile' and 'despised by all men'. Soil could perhaps fit this description, but, as we shall see, others would disagree: the matter of the first matter is not quite that simple.

This initial stage was known as the nigredo, the black stage, because it dealt with a raw, confused mass. Although modern psychologists like Jung have felt that this corresponds to the unrefined state of the unconscious before any inner work is undertaken, this interpretation of the Work may extend back as far as the semi-mythical first century adept Maria Prophetissa. Maria was one of the earliest alchemists to have understood the symbolic, or inner aspect of the work, so she may well have understood that the blackening of nigredo was equated with a spiritual or metaphorical death. In order to be cleansed by the work, the alchemist had to deal directly with the confused mass of himself or herself.

The Lesser Work: Albedo

The second stage, albedo, deals with whitening, or cleansing the matter. In physical terms, this may have involved a washing of the matter. In inward terms, we are dealing with a cleansing, perhaps referring to the observing of rituals such as fasting and other abstinences. It was hoped by such practices that the body would be readied for a reunion with the soul.

The Greater Work: Citrinitas

This stage of the work tends to disappear after the Hellenistic period, and represents a starting again, much as in nigredo, but at a purer level. It is a necessary preparation for the marriage of matter and spirit, which takes place during rubedo.

The Greater Work: Rubedo

The final stage, rubedo, is the climax of the work at which the Philosopher's Stone or Elixir is achieved. In spiritual terms, it is the mystic union of the soul and the body. This became known as the conjunctio, sometimes referred to as the chemical wedding, or the marriage of the king and queen, or sun and moon.

Unlike chemistry, the alchemist would not be conducting operations in an empirical manner; that simply wasn't the point. The work would be repeated and repeated and repeated, each time waiting for the colours to appear, each time hoping that the stars and other esoteric variables would be sympathetic to the success of the work. Each new generation or school of alchemy would add something new to the process, much the same way that a gardener may continually graft one plant onto another, and then another, as the garden grows and surrounds the gardener with riches.

Hermes Trismegistus and the Corpus Hermeticum

The most influential text in alchemy is the *Tabula Smaragdina*, or *Emerald Tablet*, ascribed to Hermes Trismegistus, who was supposed to have been a priest who introduced the Art to Egypt sometime during the Old Kingdom, but the writings ascribed to him seem to have been composed around the time of Maria Prophetissa; the oldest extant copies of these writings are found amongst the works of the great Arab alchemist Jabir, dating from the eighth century.

Hermes is the Greek messenger of the gods. In Latin, he is known as Mercurius, and Mercury, the volatile, changeable matter, plays a vital role in alchemical thinking and practice. Hermes was also identified with Thoth Hermes, the Egyptian scribe of the gods. This meant that he was naturally associated with learning, the world of spirits and with lunar cycles. Whether or not there was originally a great teacher on whom tradition later embroidered the name Hermes Trismegistus, we shall never know, but it would seem that the writings bearing his name were composed by a group of Egyptian Gnostics. The famous proverb, 'As above, so below' is from the Hermetic writings, known as the *Corpus Hermeticum*. This is actually a conflation of 'That which is above is like that which is below', which itself is taken from what is perhaps the single most important text in the entire history of alchemy. This is the *Smaragdine,* or *Emerald*, *Tablet*. It is said to have been found in Hermes' tomb, clutched in the bony hands of the departed teacher. Part of its influence derives from the fact that it is said to contain the sum of all knowledge in its dozen or so verses. The text runs as follows:

The Emerald Tablet of Hermes Trismegistus

True it is, without doubt, certain and most true. That which is above is like that which is below, and that which is below is like that which is above to accomplish the miracles of the one thing.

And as all things were made by the contemplation of one, so all things arose from this one thing by a single act of adaptation.

Its father is the Sun, its Mother is the Moon.

The Wind carried it in its womb, its nurse is the Earth.

It is the father of all miracles throughout the whole world.

Its power is perfect.

If it is cast onto Earth, it will separate the element of Earth from Fire, the subtle from the gross.

> With great sagacity it ascends gently from Earth to Heaven, and it descends again to Earth, uniting in itself the forces from above and from below. Thus you will possess the glory of the brightness of the whole world, and all obscurity will fly from you.
>
> This thing is the strength of all strengths, for it overcomes every subtle thing and penetrates every solid thing.
>
> In this way the little world was created according to the great world.
>
> In this manner marvellous adaptations will be achieved.
>
> For this reason I am called Hermes Trismegistus, because I hold the three parts of wisdom of the whole world.
>
> What I have to say about the operation of the Sun is finished.

Despite, or maybe because of, its obscurity, the *Emerald Tablet* circulated widely, and was known in Syriac, Greek, Arabic and Latin versions. Although when it was introduced to the West in the Middle Ages it was erroneously believed to be as old as history itself, it nevertheless retains something of the quality of an ur-text; all alchemists would have known it by heart. Its meaning was not necessarily to be expounded; rather it was to be lived.

In addition to this text, and next in importance in the *Corpus Hermeticum*, are the *Poimandres, or The Mind of the Sovereignty*, and the *Asclepius*. In the *Poimandres*, all of Creation is revealed to Hermes, together with the knowledge that he as a human partakes of the Divine nature; at the end of the text Hermes vows to break free of the prison of the material world to help his fellow beings reach enlightenment. The *Asclepius* describes a human being as a *magnum miraculum*, reiterating the idea that we partake, or have latent within us, Divine wonders. Also stressed is the idea that may be becoming familiar to us, that everything that exists has one thing, the One Thing of the *Emerald Tablet*, as its ultimate cause and origin.

The Symbolic Characters of Alchemy

No alchemical text ever spoke plainly, for fear of their authors being branded heretical, or of being kidnapped by the nobility (as happened frequently) and then being forced to make gold for their captors. Symbolic characters therefore play a strong role in alchemical poems and paintings, and numerous alchemists (such as Trithemius) were also pioneering code masters.

Mercury

Mercury is the most important, as he is not only related to the Art's mythical founder, Hermes Trismegistus, but is also seen as the personification of the work. Mercury means change, and that he is so important in the Art illustrates the belief that alchemy involved liberating latent forces in matter and transforming them into something greater. Indeed, alchemy is sometimes referred to as the 'art of transformation'. Mercury is either represented as a winged messenger holding a caduceus, or is sometimes portrayed as an hermaphrodite.

Lady Alchymia

She is the guiding principle of alchemy, often portrayed as a beautiful woman in flowing robes.

Old King/Young Prince

The Old King represents the matter to be worked upon, the prime matter or old state of consciousness that must be shattered during nigredo. Conversely, the Young Prince, his son, is the new matter or consciousness struggling to emerge.

Sol and Luna

Sol represents the sun and sulphur, and is the masculine principle. Luna is the moon, quicksilver, the feminine. In the laboratory, this pairing finds its expression in the alchemist working in conjunction with their partner: the *soror mystica* (mystic sister) and *frater mysterium* (secret brother). Zosimos and Theosebeia, Nicholas and Perenelle Flamel, and Thomas and Rebecca Vaughan are celebrated embodiments of the essential complementary opposites of the Great Work. From the earliest times, women were seen as equal to men, and that men and women should work together to perfect the Great Work.

The Green Lion/Red Lion

The two lions represent the beginning and end of the work. Sometimes the green lion is interchangeable with a toad, a device used by Sir George Ripley in his *Vision*.

The Egg

Usually interpreted as the retort in which the prime matter is placed. The yoke represents the gold that is waiting to be brought forth. The egg is often seen as a symbol of the soul.

The Dragon/Ouroborous

The Dragon guards the alchemical treasure. As in fairytales, it must be overcome in order for the alchemist to gain the gold. The dragon that eats its own tail, the ouroborous, signifies the cyclical nature of the work, and the harmony of nature.

The Tree

In mythology, the tree often represents the world, such as the Norse Yggdrasil. In alchemy, it usually stands for the tree of knowledge that stands in the Garden of Eden. Sometimes the tree stands in a walled garden, to gain entry to which is the alchemist's goal.

The stages of the Work have various symbolic representations: nigredo is often portrayed as a Death's Head (caput mortuum), or ravens; albedo as a king drowning, or in a heat bath, a pelican piercing its heart with its beak, a unicorn or a white rose; citrinitas as a sower sowing seeds in a field; rubedo as the wedding of the King and Queen, a starry lion, golden coins, a rose garden or the ouroborous eating its tail and arriving back at its divine origins.

The Language of the Birds

Alchemists were masters of wordplay. Many alchemical texts claim that they are 'hiding a secret openly', meaning that initiates will understand them, and that everyone else will see complete gibberish. (Gibberish, incidentally, is alchemical in origin: it was coined to describe the apparent incomprehensibility of the writings ascribed to Jabir, or Geber, as he was known in Latin.) In addition to the symbolic characters from alchemical art, such as the Green Lion and the Ouroborous, the texts frequently employ riddles, puns, and assonance. They even go to the extent of stating the exact opposite of what they are intended to convey. This wordplay is known as the Language of the Birds, or the Green Language. The great Persian poet Farid ud-din Attar's twelfth century work *The Conference of the Birds* is one of the earliest works to 'openly conceal'; he states that his work is intended to show the path of spiritual development through the journey of a group of birds to find the palace of the Simurgh, or Lord of Creation. Each bird has various skills. The Hoopoe has knowledge of Good and Evil, while the Wagtail has 'knowledge of God and the secrets of creation.'

Fulcanelli points out that before language and literacy came to use the written word so heavily, all discourse would largely be spoken, which would further illustrate the need for adepts to recognise each other through the use of coded or assonant words. This slang has always been despised (as the first matter is said to be), and is seen as ignoble, but is in fact the reverse. Such wordplay, Fulcanelli argues, should guide us towards the truth, and not leave us imprisoned in mental labyrinths of our own making.

The Importance of Number

Alchemists, from the earliest times, were obsessed with number. The art, or science, of numerology extends back to the Babylonians, and perhaps even beyond. Pythagoras stressed the importance of number, and he believed that all numbers had a mystic significance, with each number representing eternal truths, with combinations of numbers standing for the interactions between humanity and heaven. The numbers 1,2,3 and 7 are perhaps most significant, as they represent the Unity out of which all creation comes, the duality that was the first thing to be created (e.g. light and darkness in the Judeo-Christian tradition), the threefold nature of the Divine (a triple Divinity was not just restricted to Christianity), and the number of days involved in the creation of the world. Maria Prophetissa's mantra of the one becoming two, the two becoming three, the three becoming the fourth that represents the One, in essence summarises the belief - common to all mystic traditions - that the multiplicity of the world originally came from one source in which all things are unified. It was this one thing that the alchemists sought, which they, or at least the more mystically minded of them, believed was the Philosopher's Stone.

2: A Brief History of Alchemy in the West

Origins of the Word

The word alchemy itself could have originated a number of ways. It is thought to derive from *al-kimia*, a late Greek word for transmuting gold and silver. Its earliest known occurrence is in the decree issued in 296 AD by the Emperor Diocletian against the writings of the Egyptian *kimia-ists*. Suidas, writing in the ninth century, explains that the word means those who have knowledge of the Egyptian art, further enhancing Egypt's claim to be the cradle of alchemy. Suidas explains that the word is derived from the words *chemi* or *cham*, meaning Black Land, which was the ancient name for Egypt, on account of the rich, dark soil that was fed by the Nile.

The word *chymike* was first used by Alexander of Aphrodisias to denote work done in a laboratory. At the time Alexander was writing, in the third century AD, this laboratory work mainly comprised the concoction of herbal and plant remedies. The Dutch physician Herman Boerhaave (1668-1738), who made a study of alchemy in the early eighteenth century, believed it meant occult, or mystery, seeing a derivation from the Hebrew *chamaman* or *hamaman*, meaning a mystery, something that was not to be revealed to the populace at large, but should be treasured as a religious secret.

Tertullian, writing in the late second century AD, tells of how a race of giants were born of certain women who had had relations with a race of spirits. The offspring, the giants, were condemned to exile in the material world, and taught their wisdom to whomsoever asked of it. The books that held their secrets were called *chema*, and Tertullian describes them as the teachings of fallen angels. The legend was also known to the great second century theologian Clement of Alexandria, who describes them as being masters of metallurgy, biology, magic and astrology.

If we conflate these various etymologies, we are left with a word that denotes a likely Egyptian origin, something that involved laboratory procedures, something possibly of divine origin or at least with divinely inspired wisdom, and something that was secret, that was seen to be so precious as to warrant a virtual conspiracy of silence. These associations have remained with the art ever since.

In English, the art was called *alcamye* or, in Chaucer's spelling, *alcamistere*, which has variants in *alcamystere* and *alcamystrie*, which openly hint that the art was a mystery that kept its secrets.

The Prehistory of Alchemy

The exact beginnings of alchemy are unknown. As with so much of the subject, myth and reality, dreaming and waking, are intertwined; it is often impossible to ascertain whether something is 'true' or 'actually happened'. Ancient texts are often incomplete, and what survives tends either to be prosaic lists and remedies on the one hand, or impenetrable allegories and visions intended for the initiate only on the other.

The Book of Genesis is important in alchemy, as it describes the creation of the world, the emergence from the Divine of matter. Adam is frequently cited as the first alchemist, and is often regarded as something more than human, almost as one of Tertullian's fallen angels. He is the primordial man, a microcosm of the universe living in harmony with nature in Eden. The Fall is seen as Adam and Eve's exercise of free will, of their desire to immerse themselves in the physical world while still retaining a connection to the spiritual.

In tracing their lineage back to the First Man, alchemists were hoping that such an ancient lineage - none could be older, after all - would validate the art, and give it a gravity and authority that it might otherwise lack. This practice of ascribing alchemical works to figures such as Moses and Cleopatra, would continue well into the Middle Ages and Renaissance in the hope that esteemed authors would lend credence to the ideas contained in the texts.

Ancient Metalworking and Smithcraft

The origins of alchemy may lie in the ancient traditions of metalworking and smithcraft, and smith-Gods such as Vulcan were revered. Metals were believed to grow in the earth in much the same way as a baby grows in the darkness of its mother's womb. Mines were seen as sacred, as were the ores that were extracted from them. Everything in Nature was seen as alive and possessing a soul, even minerals, and it was not uncommon for the sinking of a new mine to be accompanied by religious rituals. Miners were obliged to refrain from sexual activity in the days prior to a descent to the seam being worked. To remove ores from Mother Earth was to take her precious embryos, and an unclean miner would defile the ores and thus doom any smelting done with them to failure.

Gold was mined at least as early as 3000 BC, and the discovery of smelting liberated early peoples from using tools largely as they were found. Now metals could be shaped and moulded into spears, swords and chalices. Smiths were the creators of both beautiful objects and objects of death and destruction. They were both revered as civilising heroes and as agents of the dark forces. This dual aspect was to be inherited by the alchemists.

The craft of the smith was sacred in nearly all cultures. Like the priesthood, the craft was initiatory, and marked the smith out as a person with special powers within the community. The belief that all things in the visible world have counterparts in the invisible world, the abode of spirits and gods, implies that the smith, in effecting changes in the visible, material world, was also making changes in the other world. This linked the craft inextricably with the spiritual, and suggests a possible reason for why matter and spirit were seen as being so closely aligned.

The work of the smith was to shape metals that had been taken from the womb of the Great Mother and 'improve' them through the process of smelting. It was generally assumed that metals 'grew' in the earth, and that the smith was merely aiding natural processes. It was also believed that all metals, if left in the earth, would eventually turn into gold. Again, the smith was merely speeding up a process that Nature took much longer over. Through this work, the smith was also thought to be perfecting himself. Again, this idea was to be inherited by the alchemists.

It is of some interest to note that while traditionally associated with the priesthood and the shamans, smiths were also seen as brothers to poets, musicians and dancers. Again, alchemy seems to have inherited some of this in the sense that many alchemists were also poets. Some, such as Thomas Vaughan, seem to have been alchemists solely in the poetic sense, while Michael Maier's most famous work, *Atalanta Fugiens*, is a cycle of fugues scored for three voices.

From these cloudy origins then, we can see that what was to become alchemy was a tangle of practices that involved a solitary worker, one initiated into secrets by a teacher, someone who was both feared and honoured, a person whose work was intimately bound up with nature and her sacred processes, one involved in work that was inner as well as outer, one who, although dealing with the raw materials literally clutched in handfuls from the earth, was also a brother to poets and dancers, whose work spoke of the same eternal things.

Alchemy Appears in the History Books

Alchemy is generally held to have emerged in the West in Egypt, possibly around the time of Alexander the Great (356-323 BC). However, as with much else in matters alchemical, this is strictly hypothetical. The Babylonians were known to have had a tradition of sacred metalworking, and an eighth century BC Assyrian tablet details purification rites for building a furnace. Secrets would have no doubt passed along trade routes just as readily as silk and slaves, so it is quite possible that there was a three-way flow between Babylon, Assyria and Egypt. This, the earliest school of Western alchemy, is known as the Hellenistic.

Hellenistic alchemy was shaped in part by the worldviews inherited from the Greeks. Early Greek philosophy held that all matter was derived from prime matter, the original building block of the universe. Thales of Miletus believed that all matter was essentially water; Heraclitus believed that fire was the ultimate source of all things, while for Xenophanes, all matter was derived from earth. The theory of the four elements of earth, air, fire and water seems to have been first propounded by the fifth century BC Greek philosopher Empedocles, who argued that they were governed by the twin principles of love and strife.

The Theory of the Four Elements

The theory of the four elements was developed by Aristotle (384-322 B.C.), whose writings on the elements were to remain canonical until the advent of modern empirical science in the late seventeenth century. Aristotle held that the four elements had two qualities, either hot or cold and wet or dry. Thus earth is cold and dry; air is hot and wet; fire is hot and dry; and water is cold and wet. In each element, one of the qualities predominates: in earth, it is dryness; in air, wetness; in fire, heat; and in water, coldness. The elements can change, according to properties they have in common with other elements. Fire can become air through heat, or become earth by drying out. Implicit in this is the possibility of transmutation: lead could become gold through the manipulation of the metal's qualities. (Modern physics has shown that this theory is, in essence, true, as elements can be changed into others through the manipulation of their atomic structures.)

The First Western Alchemists

The first name that comes down to us is Bolos of Mendes (c.300-c.250 BC), who wrote a book called the *Phusika kai mustika*: *phusika* refers to the things of nature, while *mustika* refers to mysticism, but it would be slightly inaccurate to translate the work as 'On Natural and Mystical Things'. Mustika has a sense in which it refers to being initiated into knowing, so the work may be more accurately rendered as *On Natural and Initiatory Things*. The concepts of nature and initiation will recur in almost all of later alchemy. Bolos seems to have been a follower of Democritus (c.460-c.370 BC), the originator of atomic theory, and for many years it was believed that the *Phusika* was in fact the work of Democritus. The book describes how a master magus dies before being able to complete his teachings, and his disciples try to contact him through necromancy in order to learn how to 'bring the natures into harmony'. The disciples are surprised when a temple column cracks open revealing a text that reads 'Nature rejoices in Nature, Nature conquers Nature, Nature masters Nature'. This was to become something of an alchemical motto, and the prominence accorded to magic, harmony and nature would be acknowledged by subsequent generations of alchemists.

Bolos' other writings cover many topics, including astrology, chemistry, agriculture, hieroglyphics and most of all, magic. His theories of magic largely centre around the ideas of antipathy and sympathy, which is to say that there are natural forces that the magician or alchemist must be aware of if they are going to have any likelihood of success in their operations. These ideas were gaining wide currency in Persia at the time, and were also adopted by various Greek schools such as the Pythagoreans, and would also be later adopted by the Neoplatonists. Bolos also seems to have been very interested in dyeing, chemistry and medical remedies, which would all play their part in later alchemy.

Hellenistic alchemy seems to have been concerned for much of its time with the dyeing and tingeing of metals. It is known that this was restricted to smiths who worked under the instruction of priests, and all work on the metals was done within the confines of temple precincts. As with prehistoric smithcraft, the Hellenistic alchemists were seen to be engaging in a sacred task, one whose secrets were not to be disclosed before the populace at large. A great deal of time was spent in the manufacture of gems and precious metals, that is to say, metals that were perceived as precious, or that may have had a greater ritual significance once they had been subjected to the heat of the furnace. Although Bolos can be dated with some degree of certainty to the third century BC, it is highly unlikely that he was the initiator of alchemy in Egypt, and the true origins of the temple metalworkers may stretch back much further.

From the Forge to the Laboratory

The next important figure after Bolos is Maria Prophetissa, otherwise known as Maria the Jewess, who may have lived in the first century BC. She is sometimes referred to as the 'Mother of Alchemy', and was a revered figure who is credited with inventing several key pieces of laboratory equipment, and originating the phrase 'One becomes two, two become three, and out of the third comes the one as the fourth'. She is credited with inventing the bains marie and the tribikos. The bains marie, or balneum mariae in Latin, is a bath that is kept warm through the simmering heat of a kettle or cauldron; it is ideal for maintaining the constant, gentle heat that alchemical writers speak of as being so important to the work. This gentle warming and washing is often associated with baptism. The tribikos is an alembic, or still, used for distillation. Maria is also thought to have been the first to stress the importance of colour in the alchemical work, and here we come to one of many areas where laboratory work and the symbolic interpretation of it seem to blur, so that it becomes impossible for non-initiates to understand the full import of the work under discussion.

What these stages refer to is at once a chemical process related to the changing colours of the matter being worked upon, and the symbolic stages that the alchemist goes through on an inner level. Whether Hellenistic alchemists such as Maria understood the process as a multi-layered, hypertextual one, is impossible to say, although Maria's writings do betray an understanding that the art was both physical and non-physical at the same time.

If Maria Prophetissa is seen as the founding mother of alchemy, then a case could be made for Zosimos of Panoplis (fl. c.300 AD) as being the founding father. Like Bolos before him, he seems to have written a great deal, including works on the obligatory subjects of agriculture, history, warfare, as well as areas closer to alchemy's heart: philosophy, medicine and chemistry. The alchemical encyclopaedia *Cheirokmeta* was written by Zosimos and his sister Theosebeia. It was written in Greek, but also used four letters from the Coptic alphabet, making 28 letters in all, illustrating that even as early as the third century, a concern for number was important to the work. Zosimos seems to have regarded the work as a discipline involving chemical operations, but whose ultimate goal was spiritual: he advised Theosebeia to 'perform these things until your soul is perfected.' Zosimos and Theosebeia can also be seen as the first Soror Mystica/Frater Mysterium pairing in alchemy.

Gnostic Influences

Zosimos (and Theosebeia too, in all probability) belonged to the Poimandres sect of Gnostics. Although there were a bewilderingly large number of Gnostic groups at this time, all with their own individual teachings, the basis underlying them all was the belief that the world was fundamentally evil, and that the purpose of a spiritual path was to free the practitioner from the prison of earthly existence. Knowledge of the Divine could not be taught by priests, but had to be experienced by the seeker themselves. This inner experience was in essence the awakening of the seeker to the realisation that the Divine, or at least a spark, was inside himself or herself, so what they were really seeking was their own True Nature. Zosimos makes clear that the work of the *magnum opus* or Great Work, should be performed on the alchemist's *terra adamica*, in other words, on the alchemist themselves, whose body is seen as the 'earth of Adam', the inheritor of the primal state.

Furthermore, Gnosticism was a heresy. Although the Western Church never declared the Art heretical, sufficient condemnations were issued to ensure that alchemy was perpetually demonised, and alchemists persecuted.

Arabic Alchemy

The fortunes of Hellenistic Egypt began to decline in the early centuries after Christ. In 387, Christian fanatics ransacked the great library at Alexandria; the Muslims destroyed forever what was left in 641. The loss of the library is one of the greatest disasters in human history: our knowledge of the ancient world would be infinitely greater had any of the texts survived. The irony of the Muslim sack of Alexandria was that almost as soon as Islam began to establish itself, around this time, it encouraged learning in a way that Christianity never did, and Islam began to inherit the wisdom of Greece and Egypt.

Tradition holds that Alchemy began in the Arab world with the Umayyad prince Khalid ibn Yazid (c.660-c.704). It is said that Khalid was the son of the Caliph Yazid I, and that, upon Yazid's death, he was succeeded by Khalid's elder brother Muawiya II. However, Muawiya died shortly after his succession, and the Caliphate should have then passed to Khalid himself, who was then still a teenager. An older relative, Murwan, was appointed instead, on the condition that Khalid would be next in line. Once he was Caliph, however, Murwan reneged on his promise, and nominated his son Abdul-Malik as his successor, and accused Khalid's mother of immorality. Khalid's mother was not a woman to be underestimated, and had Abdul-Malik killed; the story goes that he was either poisoned and/or suffocated with pillows.

These events horrified the young Khalid, and he withdrew from court life to study the sciences. In Alexandria, Khalid furthered his studies of alchemy under a Christian scholar by the name of Morienus, who is supposed to be have been a pupil of Stephanos of Alexandria (fl. 610-641), perhaps the most eminent alchemist in Egypt since the time of Zosimos. Stephanos was a favourite at the court of the Byzantine Emperor Herakleios I, who, aside from being a great military leader was also a man of learning and seems to have done much to encourage the intellectual life of the time. Like the later Holy Roman Emperor Rudolph II (1576-1612), Herakleios seems to have devoted so much time to the study of alchemy in his later years that affairs of state took second place to the search for the stone.

Morienus was said to have taught Khalid with the hope of converting him to Christianity from the new religion of Islam, and was happy to show the young prince a transmutation. Khalid was so impressed that he declared all his previous teachers to be frauds and had them all executed. Morienus took this as his cue to retire to a cave as far from Khalid as possible, and it was several years before the prince learnt where his teacher was residing. Morienus was brought back to teach once more, and this time imparted the secret to Khalid, who then set about enshrining this wisdom in verse. Among the works bearing his name are *The Book of Amulets*, *The Great and Small Books of the Scroll*, *The Book of the Testament on the Art* and *The Paradise of Wisdom*. With Khalid, we see the transmission of Hellenistic alchemy to the Islamic world, and for the next three hundred years, the Arabs become the leading exponents of the Art.

In the first centuries of its dominance, Islam actively encouraged learning, and scholars from all over the known world were invited to Damascus to translate the works of the Greek masters. Scholars who were persecuted by the Church in the West were actively welcomed, and many groups emigrated to the Arab world. (One of the most prominent groups to be involved with the translation and assimilation of Greek knowledge were the Nestorian Christians, who had been declared heretics by after the Council of Ephesus in 431 AD, clearly showing that the transmission of alchemical teachings was often an underground activity.) While Europe sank into the Dark Ages (with the exception of Ireland, where learning continued to flourish in the monasteries), the Arab countries made huge advances on all fronts. As we have seen with Khalid, the main source of transmission was the great multi-cultural crucible of Alexandria; the Mesopotamian cities of Edessa, Harran and Nisibin also participated in the traffic of knowledge that was going into the Arabic world. The result was that Arabic alchemy received the accumulated wisdom of most of the western traditions up to that time. Foremost amongst them, of course, were the Greeks, who were evidently revered above all others to judge from the number of quotations and technical terms that Arab writers borrowed from them.

Jabir: Islam's Greatest Alchemist

The first great Islamic alchemist is Jabir ibn Hayyan, known in Latin as Geber, and he is often cited as the greatest of all Arab workers in the Art. Jabir was born in the town of Tus in Khorassan (near the modern city of Meshed) around the year 721, and may have lived for some time in the city of Kufa, on the western banks of the Euphrates. His father may have been involved in the overthrow of the Umayyad dynasty by the Abbasids in 719-720; it is not certain as to whether he survived to see the birth of his son. The young Jabir was educated by Bedouin, and also seems to have been a member of the Sufis, the mystical branch of Islam that rejected the luxuries of court life for an austere life of prayer, contemplation and ecstasy.

Jabir's father's connection with court life, however, may have secured his son a position as resident alchemist to the sixth Shi'ite Imam Ja'far al-Sadiq (700-765). Jabir wrote extensively, and may have personally overseen the translation of numerous works into Arabic.

Apart from his alchemical writings, among his works can be numbered books on astronomy, philosophy, logic, medicine, and warfare; he also wrote on the more esoteric topics of automata, magic squares, and mirrors in addition to commentaries on Euclid, Ptolemy and Apollonius of Tyana. He is said to have died in 815, either in his laboratory in Kufa or in his birthplace of Tus, with a copy of his *Book of Mercy* under his pillow; some authorities believe he survived well into the rule of Al-Ma'mum (813-833).

Jabir's importance lies in the scope of his work, and the thoroughness of his approach. Although he may have been a Sufi, his work was not tainted by the sort of mystical speculation that had thrived in Alexandria. With Jabir, we can see the beginnings of what we would now call the scientific method. He developed a theory that all metals were generated by the interaction of sulphur and mercury, and grew in the earth under the influence of the planets. The planets had been associated with metals from the time of the first Babylonian astrologers, with Saturn governing lead, Mars iron, Jupiter tin, Venus copper, Mercury mercury, the Moon silver and the Sun gold. He was also obsessed with numerology, and seems to have developed the ideas of the Pythagoreans. For Jabir, transmutation was a precise art governed by number, and the study of numbers could help the alchemist achieve his goal. And the earliest surviving copies of *The Emerald Tablet* come from Jabir's library.

The Origins of European Alchemy: The Translators

Alchemy's introduction into the West seems to have been the result of various disparate forces. After the First Crusade (1095-1099), Christianity began to have more contact with Islam, and found that Muslim science was far in advance of the West. Spain was still under Moorish control, yet this was the most fertile area of cross pollenisation. Students were welcomed at the great universities of Toledo, Barcelona, Segovia and Pamplona regardless of religious beliefs, and it was in Spain that the first translations of alchemical books to reach the west were made.

Robert of Chester, who, despite his name, seems to have come from Rutland, was studying alchemy and astrology with his friend Hermann the Dalmatian when he was approached by the Abbot of Cluny, Peter the Venerable, with the suggestion of translating the Koran into Latin.

After completing this, in 1143, Robert began work on translating Morienus' *The Book of the Composition of Alchemy*. He completed this on 11 February 1144, making it the first alchemical work to appear in Latin in Europe, and giving Morienus a part to play in the origins of Western alchemy in addition to Islamic. This translation ushered in a veritable craze to translate Arab authors, and over the next hundred or so years, a vast body of work found its way into Europe. Robert himself continued working, translating the work of the mathematician Al-Khwarizmi, and introducing algebra to the West in the process.

Just as various groups had formed to translate the Greeks into Arabic four hundred years earlier, so the process was repeated in the West. The Archbishop of Toledo founded a college devoted solely to translating, and it was here that the greatest of the Latin translators worked. Gerard of Cremona (c.1114-1187) is thought to have translated 76 texts, including works by Avicenna, Aristotle, Ptolemy and Jabir.

Jabir's posthumous career as a writer in Latin is an interesting one. Latinized, he becomes Geber (the G is soft), and manages to continue producing works more than three hundred years after his death. Although Gerard of Cremona can be excused, it seems that other translators actually started to write works and attribute them to Jabir. This has a long tradition, as we have seen. Geber's works differ from Jabir's, in that such Jabirian preoccupations as balance and number are conspicuous by their absence. The texts attributed to Geber were enormously influential, among the most notable being *The Summit of Perfection, The Investigation of Perfection* and *The Discovery of Truth.*

Working with the translators were men such as Bartholomew the Englishman (fl.1230s) and Vincent of Beauvais (c.1190-1264), who edited the texts and wrote commentaries on them. The thirteenth century sees alchemy entering the mainstream of European thought, perhaps the only time that it did so. All of the great minds of the period had something to say on alchemy, and here we start to detect the presence of European adepts who were starting to write original works themselves, and not merely reworking the Arabs.

The Renaissance of the Twelfth Century

As we have seen with the Greeks and the Arabs, alchemy tended to attract scholars and priests. The reasons for this are simple: learning was usually restricted to the clergy, with most people being more concerned with survival than literacy. In order to read, you usually had to belong to a religious order, and monasteries became hotbeds of alchemical experimentation. Both Bartholomew and Vincent were monks, and the first translations into Latin were carried out under the auspices of a bishop.

Although the church in the West never welcomed learning in the way that the Arabs did, or the Byzantines under the great scholar Michael Psellus (1018-c.1078), who offered a free university education to anyone who wanted it, there was, for a brief period, a climate of enquiry that put alchemy firmly on the map in Europe. Although one may associate alchemy with mediaeval ignorance, explosions of interest in it tended to occur in times of flourishing intellectual enquiry. Just as Hellenistic alchemy flourished in the vibrant atmosphere of Alexandria in its heyday, so the West embraced alchemy in an atmosphere of renewed study, the so-called Renaissance of the twelfth century.

Most of Europe at the time the first translations were made was dominated by the Papacy and the Holy Roman Empire, who spent much of the century locked in conflict. Cities were expanding, and Europe's economy was slowly becoming both urbanized and rationalized. Advances made in many walks of life - in navigation, animal husbandry and Fibonacci's new method of double-entry bookkeeping - constituted a mini Renaissance. Intellectual life was slowly responding to these changes. Official learning was dominated by the legacy of the University of Paris, where logic and the still only partially translated works of Aristotle held sway. The work of Arab scholars such as Avicenna (980-1037) and Averroes (1126-98) was breaking new ground; indeed the advances of the Arabs provided crucial fuel for the renaissance that Europe was experiencing at this time.

The First Western Adepts

The first European masters were not exclusively alchemists. Nearly all of the great names of the thirteenth century were scholars who devoted themselves to the advancement of learning in all its forms. The first name in European alchemy is that of Albertus Magnus (1193-1280), who was born in Lauingen in Swabia, and became a monk in 1223 when he joined the Dominicans in Padua. Although apparently not very tall, he became known as Albert the Great, the *Doctor Universalis* who excelled in all branches of learning. From 1228 to 1245 he taught at Freiburg, Ratisbon, Strasbourg and Cologne, before moving to Paris, where he taught St Thomas Aquinas and may have met Roger Bacon.

Like Jabir before him and Leonardo da Vinci after, Roger Bacon was a great polymath whose mind seems unfettered by the time he lived in. He was frequently accused of sorcery, and may have endured a period of imprisonment. His work stands at the meeting place of Medieval and Modern Europe, and his ideas anticipated both the Renaissance and the Enlightenment; he was lauded as a scientist and thinker yet branded a wizard. Some of his inventions had to wait centuries for realisation.

Although seen as the first champion of the scientific method, he was also rumoured to have performed alchemical transmutations, and remains an enigmatic character, still steeped in mystery and legend. Like Pope Sylvester II, he was rumoured to have a talking head made of brass, and it was said that he gained his extraordinary learning from the Devil. He is supposed to have drawn up plans for flying machines and machines that enabled a man to breathe at the bottom of the ocean, and to have constructed a mirror that enabled him to see far-off events. He is also said to have invented a microscope, a tank and a pontoon bridge. When he died, his books were nailed to the shelves they sat on and left to rot, such was the conviction felt by his fellow Franciscans that he had been in league with the forces of darkness.

Bacon was born at Ilchester in Somerset around the year 1214 to a wealthy family who supported Henry III in the war against the Barons (a position that would later drive them to ruin). He probably was sent up to Oxford at the traditional age of ten or twelve, and there proved himself to be an exceptional student, being taught by the most learned men of the day, including Robert Grosseteste, Bishop of Lincoln and the leading mathematician of the time. Perhaps while an undergraduate, Bacon became a Franciscan.

After Oxford, Bacon went to Paris, which boasted the greatest university of the day. Bacon's tutor at Paris, unlike Oxford, was an obscure solitary by the name of Master Peter, a shadowy figure who has escaped the history books almost completely. Whereas, in Oxford, Bacon had received the standard university education of the time, in Paris, he was instructed in Peter's own unique style. He learnt alchemy and astrology, but was also instructed in the empirical observation of nature: Bacon was to refer to Master Peter as the 'Lord of Experimentation'. Bacon had already been introduced to this method by Grosseteste, and would later develop the work of both his masters, and in doing so, lay the foundation for modern science.

Bacon came to believe that through studying Nature, Man could come to have knowledge of the Creator. He returned to Oxford a Doctor of Divinity, and began to undertake a series of experiments. For most of the 1250s, he seems to have been engaged in diverse works, including astronomy (his celebrated observation tower survived well into the eighteenth century), botany, chemistry, medicine, optics and mathematics. He gathered a small group of students around him - including Friar Bungay, another doctor of Divinity and supposed sorcerer - and set rumour mongers into overdrive with his stargazing and strange instruments. This period was the happiest of Bacon's life.

Bacon's belief that the principles of mathematics and number underlined everything in the world led to his discovery that the calendar was in need of reform; his suggestions were finally accepted in 1592 as the Gregorian system, as it was known, was finally adopted. It also led to his creating one of the earliest maps of the world. (Now rumoured to be in the Vatican library.)

But this was not all. Bacon saw abuse and ignorance everywhere, and railed against fellow scholars (including Thomas Aquinas and Albert the Great). Heretically, he judged the world of Classical Antiquity to be morally superior to that of the Christians, and called for greater stringency in Universities. He himself learned Hebrew, Chaldean and Greek in an attempt to try and get both the Bible and the Classical authors translated properly. Corrupt texts, he reasoned, would lead to corrupt students. He also believed that great secrets were contained in words: if, as the Fourth Gospel asserts, 'In the Beginning was the Word', then a greater knowledge of linguistics would lead to a greater knowledge of God. Bacon must therefore have studied what alchemists have called the Language of the Birds, the derivation of occult secrets through the use of etymology and wordplay.

Bacon's outbursts, which would continue unabated into old age, together with the popular perception that he was a necromancer, led to his summons to a Franciscan kangaroo court in 1257. It is said that he was placed under house arrest for ten years, although he himself claimed that he withdrew from Oxford life for some years due to ill health. Perhaps Bacon realised that he was sailing close to the wind, and should lie low until either his reputation improved, or a powerful patron should appear.

In 1263, that is precisely what happened. The newly appointed Papal legate to England, Guido Fulcode, heard of Bacon through an intermediary, one Raymond de Laon. The monk of Oxford, Fulcode learned, was possessed of wonderful secrets, and the legate determined to correspond with him. By the time the two made contact, Fulcode had become Pope Clement IV, and, once he deemed it safe to do so, had Bacon released in 1267 on the condition that he wrote all his discoveries down into one book.

Bacon did not write one book for the Pope. He was to write five. The *Opus Majus* was the first of these, and it is his masterpiece. In it, he followed the fashion of the time for trying to encapsulate all human knowledge in one book. It is a vast compendium surveying the sciences as they stood in 1267, and as they stood according to Bacon, who believed that all branches of science were connected; in effect, it is a medieval quest for a 'Theory of Everything'. (Another, later alchemist, Sir Isaac Newton, also dreamed of such a theory.)

Bacon hoped that in producing the *Opus Majus*, he could persuade the Pope into both offering him protection from his persecutors, and into introducing reforms in both Church and Society. The plan was scuppered in 1269 with Clement's death. Back in Oxford and reunited with his students, Bacon produced three more works, the *Opus Minus*, the *Opus Tertium* and a treatise whose title has been lost, for the new Pope, Gregory X, before falling foul of the Franciscan hierarchies again. Bacon's supporters tried to save him, but in 1278 he was imprisoned for fourteen years. It is said that he was only released when he revealed certain alchemical secrets to the head of the Franciscans, Raymond Gaufredi. If his opponents had hoped that jail and old age would have mellowed Bacon, they were to hope in vain. Upon his release, Bacon immediately set to work on another book, the *Compendium Theologiae*, an attack on what he saw as the theological errors of the time.

St Thomas Aquinas is more commonly known as the leading Scholastic philosopher of the thirteenth century, and was Albertus Magnus' star pupil. Like his master and Roger Bacon, he devoted some time to the ideas behind alchemy, and is held by some to be the author of *Aurora Consurgens*, an alchemical reading of the Song of Songs.

Aquinas had been working on what was intended to be his masterpiece, the *Summa Theologica*. After four years of monumental productivity in Paris and finally in Naples - during which he was writing not only the *Summa*, but also the commentaries on Aristotle as well as lecturing - he underwent a mystical experience while saying mass on

December 6, 1273. He told his secretaries that 'all I have written seems to me like so much straw compared with what has been revealed to me.' He halted work on Part Three of the *Summa* and never wrote anything again. In the early part of the New Year, he was invited to attend a Church council in Lyons, but was taken ill on the way on March 7 1274 at Santa Maria di Fossa Nuova. (He is reported to have collided with the overhanging branch of a tree, which knocked him from his donkey.) The monks, having heard of Father Thomas' great reputation, begged him for a teaching. Aquinas, despite being not at all well, obliged with a seminar on the Song of Songs.

The *Aurora*, as it has come down to us, is the description of a life-shattering experience that could very well be Aquinas' revelations of December 1273. It is certainly possible that it was a transcription of Aquinas' final talk, as other of his works as we know them today were also transcribed by pupils. The work describes a vision of Sophia, the Wisdom of God. Aquinas realises he has neglected the Feminine Principle all his life, and she has revealed herself to him in all Her Glory. Perhaps this is what he meant when he said that everything he'd done seemed worthless, like straw in comparison. Her wisdom is equated with the Philosopher's Stone, and links alchemy with the Feminine face of God. Had he lived, Aquinas may well have found himself in trouble for his views: his work was denounced in the Paris Condemnations of 1277, which also attacked Bacon.

Alchemical Manuscripts and Images

Aurora Consurgens began to circulate in manuscript form sometime during the fourteenth century. Before the invention of printing, alchemical ideas circulated in manuscript form, and some, such as the *Aurora*, proved to be highly influential. Indeed, the manuscript tradition was essential to the dissemination of alchemical ideas, and no doubt there was a black market trade in manuscripts.

The best known version of the *Aurora* is profusely illustrated, showing scenes of traditional alchemical figures such as the hermaphrodite, the dragon and the ouroborous, the worm that eats its own tail, together with more Bosch-like creatures such as human-monkey hybrids, and sun and moon-headed knights. In perhaps what is the most shocking image, two naked figures sit side by side. The top of the figure on the right's head has been peeled back like a tin can, and also has an incision in its chest where it has seemingly reached inside its own body to take out its heart, and is offering it to the other figure, who is eating or drinking from a cup.

Although strong and verging on heresy, these images, like that of alchemical poems, were intended to openly conceal the secret in works that could only be understood by the initiated, and that, as we have seen, could only be done through contact with a genuine master. To the uninitiated, all they would see would be knights - both male and female - engaged in battles with fabulous creatures and not have the slightest idea of their great worth.

The Spread of Alchemy

By the time the *Aurora* began to circulate in the late fourteenth century, alchemy had become widespread in the monastic world, as Thomas Norton, writing a hundred years later, testifies:

> Popes with Cardinalls of Dignity,
>
> Archbyshopes with Byshopes of high degree,
>
> With Abbots and Priors of Religion,
>
> With Friars, Heremites, and Preests manie one,
>
> And Kings with Princes and Lords great of blood...
>
> For every estate desireth after good.

Norton notes that not only were bishops and princes 'desireth after good', but also:

> Men of all classes desire to partake of our good things:
>
> Merchants, and those who exercise their craft at the forge,
>
> Are led captive by a longing to know this Art;
>
> Nor are common mechanics content to be excluded from a share in it:
>
> They love the Art as dearly as great lords.
>
> The goldsmiths are consumed with the desire of knowing-
>
> Though them we may excuse since they have
>
> Daily before their eyes that which they long to possess.
>
> But we may wonder that weavers, freemasons, tailors,
>
> Cobblers and needy priests join in the general
>
> Search after the Philosopher's Stone.

Frauds naturally abounded, and Norton notes that 'many of these workmen, however/ have been deceived by giving credulous heed to impostors/ who helped them convert their gold into smoke.' This echoes Chaucer, who writes of a fraudulent alchemical priest in *The Canon's Yeoman's Tale* (c.1390).

So great was the craze for alchemical transmutations, that decrees were issued against it all over Europe. As we have seen, alchemy was banned as early as 144 BC in China; Pope John XXII forbade the manufacture of gold by alchemical means in his Bull of 1317 (but was then said to have created a fortune of his own under the direction of the alchemist John Dastyn). In England, King Henry IV outlawed alchemy in 1403; thereafter the Art could only be practiced under Royal licence. Part of the reason for this widespread outlawing was simple forgery.

There are numerous stories about alchemical coinage being circulated, perhaps the earliest being the apocryphal account of Edward II. The story holds that King Edward wanted to make gold, and charged the Majorcan scholar Raymond Lully (c.1235-1316) with the task. Lully set up his laboratory in the Tower of London on the condition that the King use the gold solely for financing his next Crusade against the Turk. The King broke his promise, and Lully is said to have escaped to France. Edward is supposed to have struck coins, called Rose Nobles, from the gold, which were still circulating in 1696, when Isaac Newton took over as Warden of the Royal Mint and began his reformation of British currency.

Another reason, lurking behind the obvious threat that a debased coinage holds to any society, is the threat that alchemy posed to the Church. After the great climate of enquiry in the thirteenth century, the Church began to become increasingly divided, with rival popes based in Rome and Avignon. With such division, there was perhaps a greater need for orthodoxy to keep the Church from imploding altogether, and anything heterodox was immediately suspect. Even as early as the late thirteenth century, figures suspected of deviating from Church dogma were declared heretical. In the Bull of 1277, Pope John XXI condemned both Thomas Aquinas and Roger Bacon, amongst many others; Aquinas was safely dead, but Bacon found himself imprisoned for fourteen years on account of his work. (Even what we would nowadays call science, let alone alchemy, was believed to be black magic, and Bacon would have no doubt been imprisoned for the totality of his beliefs, not just his alchemical experimentation.)

Nevertheless, despite widespread condemnation, serious students of the Art continued to flourish. One of the most celebrated is the French alchemist Nicholas Flamel (1330-?1417), whose story encapsulates several key aspects of what was then, as now, seen as 'real' alchemy. Flamel was not a monk, but a scrivener, who earned his living from copying documents in Paris. These could be anything from books to deeds of covenant or letters, and he would have no doubt encountered alchemical manuscripts in the course of his work. One night an angel came to him in a dream and showed him an old book with magnificent illuminations. Flamel reached out to touch the book, but both the book and the angel vanished before he could hold it in his hands. Sometime later, in 1357, Flamel bought a book and at once recognised it as the book that had been shown to him in his dream. The book was written by a certain Abraham Eleazar, and it appeared that he was describing the art of transmuting metals into gold. On every seventh page there were enigmatic illustrations purporting to show part of the process. Flamel confided in his wife, Perenelle, that he could not understand more than the first few pages of the book.

After twenty years of study and failed experiments, Flamel decided to make a pilgrimage to Santiago de Compostela, in the hope that he might find a learned Jew to explain the work to him (Abraham having written his book for fellow Jews). He went there in 1378, and spent more than a year there before meeting with one Master Canches, a merchant, who explained the figures for him, realizing that Flamel had in his possession an ancient Kabbalistic text that was thought to have been lost forever. Canches returned to France with Flamel, eager to see the original from which Flamel had made his copies, but died before reaching Paris.

Upon his return, Flamel worked without ceasing for four years, until, on Monday 17 January 1382, at around noon, he is said to have completed the work, with Perenelle as his only witness. Flamel and Perenelle are the archetypal male and female alchemists working together in harmony, the frater mysterium and soror mystica bringing complementary opposites - male and female, yin and yang - together for the purposes of the Work. (Another aspect of alchemy that would have been frowned upon by the Church, no doubt.) On April 25 that year, they repeated the success of the first experiment, and the Flamels suddenly began to acquire huge wealth. Over the next fifteen years, they founded fourteen hospitals in Paris alone, gave substantial donations to seven churches, and gave similar gifts in Boulogne, which could have been Perenelle's home town.

Such lavish gifts were seen by contemporaries as being the result of the pair's success in the laboratory, and upon Flamel's death in March 1417, a mob ransacked the house to find the source of their treasure. They found nothing, but, interestingly, both Flamel's tomb and that of Perenelle's were found to be empty when they too were broken open. They were said to have gone to India, where they were seen around the year 1700, and were reported to have been seen at the Paris Opera in 1761.

Even discounting this part of the story for a moment, it is interesting to note that all the supposedly 'genuine' alchemists lived, on average, twice as long as the average span: Jabir lived to around his mid 90s, Albertus Magnus to 87, Roger Bacon to around 80, Lully to 81, Flamel himself to at the very least 87, and, later, Isaac Newton to 84, and Fulcanelli to well over 100, according to some accounts. What enabled them to attain such lifespans, which, even now, are impressive? Could secrets have come from the East? As we shall see in the next chapter, Eastern alchemy was primarily concerned with the prolongation of human life, and India boasts a long history of 'immortals', sages who have exceeded the natural span of human life through various ascetic practices.

Alchemy in England in the Later Middle Ages: Ripley, Norton and Charnock

Alchemy in England at this time is comparatively well documented, and it contains a number of names that have endured as some of this country's greatest exponents of the Art. The period is also interesting, as it affords us a glimpse of how the tradition was passed on from one alchemist to the next.

The most prominent alchemist in England around this time was Sir George Ripley, who was the Canon of Bridlington Priory in Yorkshire. He studied on the Continent, reading theology at Rome and Louvain. He travelled widely, and at one point reached Rhodes, where he was said to have been initiated into the hermetic arts by the Knights of St John, who, like the Templars, were often accused of being sorcerers and heretics. Such was Sir George's success in the laboratory, that he donated the then astronomical sum of £100,000 a year to the Knights in their crusades against the infidel.

With a standing order to the Knights of St John seemingly well in place, Ripley returned to England, where he became Canon at Bridlington. Oblivious to the fact that the fumes emanating from his laboratory were causing concern among the other monks, he produced his best-known work, the *Compound of Alchymie* (1471), which he dedicated to King Edward IV (who, as we have seen, was so interested in alchemy as to kidnap the unfortunate Thomas Daulton).

Despite the fumes bringing attention to his work in Bridlington, Ripley managed to reach old age unmolested by kings or popes. His stock was high in his lifetime, and he was highly regarded for hundreds of years after his death in 1490. He is said to have passed the secret on to a William Holloweye, the last Prior of Bath Abbey, and also to a certain Canon of Lichfield, who in turn, transmitted it to Thomas Norton.

Thomas Norton is best known for his long alchemical poem, *The Ordinall of Alchemy* (1477), which is subtitled *Crede-Mihi*, or Believe-Me, by which Norton, in typical alchemical fashion, implies that his work, and no other, contains the secret. He is known to have been born in Bristol, and, like Flamel, was a layman. His father was, at various times, sheriff, mayor and MP for Bristol, and Thomas himself may have been either a successful merchant, or a privy councillor to that alchemically inclined monarch, Edward IV. A Thomas Norton is known to have accused the Mayor of Bristol of treason in 1478, and is said to have challenged him to a duel in the council chamber. Whether this is the alchemist Thomas Norton is unclear, but given Norton's connections to the office of mayor, it is quite likely.

In the *Ordinall,* Norton speaks of having to ride for over 100 days to find a master with whom to study. Some believe that this was Ripley, while other traditions hold that it was the mysterious 'Canon of Lichfield'. Norton succeeded in the Great Work at the comparatively young age of twenty-eight. Disaster struck, however, when his elixir was stolen by a laboratory assistant. He began the work again, only this time to have the elixir stolen by the wife of master mason William Canynges, who used the financial gain that possessing the Stone brought to build of one of Bristol's most celebrated churches, St Mary Redcliffe.

Much is known about the life of Thomas Charnock, as he wrote an autobiography in verse detailing his pursuit of the stone. He was not a churchman, being an 'unlettered scholar'. Charnock was born in Kent, either in Faversham or the Isle of Thanet, probably in the year 1524. Charnock was a laboratory assistant to an alchemist named James Sauler, who passed on the secret to Charnock on his deathbed in 1554. Charnock suffered a major setback on New Year's Day 1555, when a fire destroyed his laboratory. His fortunes revived briefly when he was taught by one of Ripley's pupils, William Holloweye, the last prior of Bath Abbey before the dissolution of the monasteries.

Holloweye had apparently made the stone, and had hidden the red medicine in a cranny in a wall when the king's men arrived to start smashing the place up. Apparently they found the red powder, and, believing it to be of no great worth, threw it onto a dung heap. The powder proved to be a great fertiliser, however, and when the dung was used in field of corn, it produced and abundant crop. The loss of the red medicine temporarily drove Holloweye into some sort of breakdown, but he seems to have recovered his wits by 1555, when he is said to have instructed Charnock.

Charnock's fortunes once again swung back the other way, when he was forced to serve in the army defending Calais. Charnock determined that no one should learn of the secret while he was away, and chose to destroy his laboratory with a hatchet. After serving in Calais, he returned to Britain, married a woman called Agnes Norton from (Stockland?) Bristol and set up a third laboratory in the village of Combwich, near Bridgwater in Somerset. His wife may well have been related (a grand daughter?) to Thomas Norton. He is supposed to have succeeded in the Work in 1574. The fumes bellowing forth from his chimney were a constant nuisance to his neighbours, and, following his death in April 1581, the house fell into disrepair, and seems to have been regarded as something of a haunted house, shunned by the villagers. Today nothing remains of the house where Charnock lived, although the lane where his house stood is still used.

Alchemy in the Renaissance

In 1452, Constantinople fell to the Turks. This precipitated a huge exodus of scholars and priests, many carrying priceless manuscripts that had never been seen in the West. Amongst the texts that arrived in Europe was the *Corpus Hermeticum*, which was translated into Italian by Marsilio Ficino at the behest of Cosimo de Medici. Cosimo, apart from being the Pope's banker, was passionately interested in the new learning that was sweeping Europe, and urged Ficino to abandon his then work-in-progress, a translation of the complete works of Plato, in favour of the Hermetic works. Cosimo had become interested in Hermeticism during the Council of Florence (1438-39), at which the Orthodox and Catholic churches tried to iron out their differences. Amongst the Byzantine scholars who attended was one George Gemistus, aka Plethon, who was effectively preaching a return to paganism. (Plethon, incidentally, is another near-centenarian.)

While alchemists such as George Ripley and Thomas Norton continued to work in the seclusion of their laboratories in the second half of the fifteenth century, Europe began to experience the beginnings of the Renaissance. Hellenistic ideas played a central role in this, and Ficino's translations made the teachings of Hermes Trismegistus available in Europe for the first time. Reformers of all kinds began to flourish, with thinkers such as Ficino and Pico della Mirandola practicing natural (i.e. white) magic and integrating Hermetic and Hellenistic ideas into their writings.

This found expression in England in the work of John Dee (1527-1608), magician, diplomat, cartographer and Elizabeth I's personal astrologer. Dee was perhaps the original architect of the British Empire, although not the empire as history knows it: Dee's vision was of a utopia based on ideals derived from antiquity, an earlier version of William Blake's vision of Albion. Dee had contact with many like-minded individuals, and it seems that, behind the scenes, efforts were made to bring about a sort of alchemical renaissance.

Alchemy had its own reformer in the shape of Philippus Aureolus Theophrastus Bombastus von Hohenheim, who adopted the nom de plume of Paracelsus. Paracelsus was born on 17 December 1493 south of Zurich, and was encouraged to practice medicine by his father. He studied at the University of Basle between 1509 and 1513, and then seems to have led a peripatetic life, travelling through Germany, France, Belgium, England, Scandinavia and Italy. He may have even got as far as Russia and the East. He may have been acting as a military surgeon during some or most of this time, and returned to Basle in 1526 where, at the behest of Erasmus, he became town physician.

Paracelsus, in one respect, was the father of holistic medicine, believing that to treat an illness, one had to treat the whole person. For Paracelsus, the doctrine of the macrocosm/microcosm was a literal truth. He was also a staunch believer in the power of the mind in determining the success of a patient's recovery. In Basle he was able to propagate his revolutionary ideas about medicine - largely derived from alchemy - but soon fell foul of the town elders when he publicly burned the works of such hallowed medical authorities as Galen and Hippocrates. Always a larger than life figure, he left Basle in 1528 unrepentant, and spent the remaining years of his life travelling and writing reams of iconoclastic works in which he extolled the virtues of medicines made from minerals (as opposed to herbs), and reiterating his belief in the astral influence upon human health. Paracelsus believed that there were heavens within us, too, and during illness, these needed to be realigned with the greater heavens in order to restore health. He died in Salzburg in 1541, possibly poisoned by his enemies.

The importance of Paracelsus lies not just in the fact that he was the first homeopath in Western Europe, but also in that he understood certain physical functions to be simply chemical reactions, that could be treated through the application of certain other chemically prepared remedies. This makes him also one of the founding fathers of modern allotropic medicine. A century before it was officially recognized, Paracelsus knew of the circulation of the blood. Here we have the classic instance, like Roger Bacon before him, of an alchemist making discoveries that were as yet unknown in the wider world, discoveries that would have to wait decades, sometimes even centuries, to be acknowledged. In Paracelsus, alchemy is taken out of the laboratory and used expressly in the service of others. His work also provides an interesting link between Western and Eastern alchemy, which we shall look at in the next chapter.

Alchemy and Music

Alchemical and hermetic ideas flourished in the Renaissance. The Florentine Camerata, set up along the lines of Ficino's Florentine Academy of a hundred years earlier, is one of numerous examples of Renaissance thinkers actively trying to incorporate the ancient wisdom into their works. The leading lights of the Camerata were composers such as Jacopo Peri and Giulio Caccini, and it was Peri who composed the first opera, *Dafne*, in 1598, and, two years later, *Euridice* (which, unlike the earlier work, has survived complete). In both of these works, music is linked with the salvation of the soul, and all the early operas were self-referential in that they were artworks that took art as their subject. The story of Eurydice was later set to music in 1609 by opera's first genius, Claudio Monteverdi, who was also a practicing alchemist. His first patron, Vincenzo Gonzaga, initiated the young composer into the Art, and Monteverdi, by his own admission, spent his life trying to embody the eternal truths in his music.

Monteverdi was not the only alchemical musician, however. Ficino, although more a magician than an alchemist, had composed hymns to Orpheus that were designed to be sung at certain times and in certain places, usually out of doors; and the German alchemist Michael Maier's (1578-1622) most celebrated work is *Atalanta Fugiens* of 1617, where each alchemical poem has both an illustration corresponding to the relevant part of the work, and the music to which the poem is set.

Alchemy and the Rise of Science

The new learning of the Renaissance slowly began to evolve into the beginnings of modern science, and the modern worldview. Two of the figures instrumental in this were Francis Bacon and Rene Descartes. Like Roger Bacon, Francis extolled the experimental method, but, unlike Roger, stressed analysis rather than synthesis, thus beginning the fragmentation of reality that has dominated the West ever since. Descartes, in the eyes of the alchemists, committed an even worse sin, that of universal doubt. Whereas the genuine alchemist would have universal, or unwavering, faith in both his/her work and the world (and also the higher world that it reflected), Descartes famously achieved his breakthrough when he resolved to doubt everything in the world, until he came to doubt doubt itself. Realizing that if he was doubting doubt, then it must mean he - or something - existed in order to do the doubting, he produced his famous maxim Cogito Ergo Sum, I Think Therefore I am (by which he really meant, I Doubt, Therefore I am). Although such developments were understandable attempts to shake off the cloak of superstition that had dominated mediaeval attempts at fathoming the mysteries of the universe, they inadvertently caused a paradigm shift in which harmony, one of the guiding principles of all human endeavour up till that point, was lost.

As these beliefs slowly gained currency, alchemy also underwent a change. From the early seventeenth century, it became increasingly mystical. Writers such as Jakob Boehme (1575-1624) and Thomas Vaughan (1621-1665) used alchemical imagery in their writings, but were not thought to have been laboratory workers. (Thomas Vaughan apparently dipped his toe in the water, but later described laboratory work as 'the torture of metals'.) These writers, Boehme in particular, equated the Philosopher's Stone with Christ and, in doing so, expressly Christianised European alchemy.

A further development in the seventeenth century was the proliferation of alchemical books. Apart from Michael Maier's *Atalanta Fugiens*, other celebrated works of the period included the *Book of Lambspring* (1599), Heinrich Khunrath's *Ampitheatre of Eternal Wisdom* (1609), and the anonymous *Mutus Liber* (1677), an illustrated work that entirely dispensed with a written text. Earlier works were gathered together and issued as large anthologies, such as the *Theatrum Chemicum* (1602), and the *Theatrum Chemicum Britannicum* (1652), edited by Elias Ashmole.

While alchemy, going further underground than ever before, continued in its emphasis on subjective experience, the need for analysis free from subjective beliefs became the guiding principle of the first scientists. In England, there were groups meeting in Oxford and London, which comprised the prominent thinkers of the day, including such men as Ashmole, Robert Boyle and Christopher Wren. Their so-called 'Invisible College' became the Royal Society in 1660, and the seal of Royal patronage meant that from now on, the prominence of alchemy would gradually diminish as research become ever more based in the empiricism championed by Francis Bacon and Descartes.

However, that is not to say the Art was no longer practiced: Charles II had an alchemical laboratory set up beneath the royal bedchamber, and Robert Boyle, although publicly repudiating alchemy in his book *The Sceptical Chymist*, still practiced the Art in secret. And in 1672, there was elected to the Royal Society a man who has come to be regarded as the greatest scientist of all time, yet who was also a practising alchemist.

Sir Isaac Newton regarded his work in the laboratory to be of more importance than his work in mathematics, which spawned the theory of gravity. Indeed, it is possible that Newton's alchemical researches directly inspired his greatest work, the *Principia* of 1687, in which his theories were first laid out. He clashed furiously with Boyle, who published his alchemical findings without compunction; Newton, however, regarded the secrets of alchemy as not only precious, but actively dangerous if they fell into the wrong hands. To see Newton as simply misled is to seriously misunderstand him. He did not have a fragmented view of reality, but he saw it as a giant riddle posed by God, one that he felt almost divinely appointed (modesty was not the greatest of his virtues) to solve. After his death, Newton's friends covered up his alchemical interests so well, that it was not until 1936 that evidence came to light when a collection of his papers came up for auction at Sotheby's. Although Newton almost single-handedly created the modern world, it is something of an irony to recall that the man whom most scientists regard as a godlike figure passionately believed in what to them would be anathema. Like Einstein after him, Newton, saw the divine fingerprints everywhere, and worked without ceasing to uncover their meaning.

The Artist Elias

Although the scientific method as we now know it was beginning to get a firm foothold on the European intelligentsia, this period also sees what is perhaps the most celebrated (after the Flamels), and certainly most detailed, account of transmutation. It was published in 1667 by the great Swiss doctor Helvetius, personal physician to the Prince of Orange, and one of the leading names in the medicine of the time.

Helvetius records that on December 27, 1666, a stranger called at his house in The Hague. He describes the man as about 42 years of age, dark-haired, and who did not wipe the snow from his boots as he stepped in from the street. Beneath his cloak he wore several gold medals, each one inscribed with verses like 'Speak not of God without Light' and 'God, Nature and the Spagyric Art [alchemy] make nothing in vain'. The man introduced himself as brass-founder named Elias, and informed Helvetius that he was a little concerned over a tract Helvetius had written denouncing the work of the English alchemist Sir Kenelm Digby. He asked Helvetius if he believed that the Philosopher's Stone existed. Helvetius, naturally enough, expressed his doubts, at which point the stranger produced three small yellow stones from an ivory box. He told Helvetius that he was looking at the Philosopher's Stone. At this point, Helvetius seems to have become intrigued, and asked Elias if he could perform a transmutation. Elias declined, on the grounds that it was not the right time.

The stranger returned three weeks later, and gave Helvetius one of the stones. Helvetius commented that it was so small it didn't look particularly impressive, at which point the stranger broke the stone in half and threw half on the fire, much to Helvetius' horror. He gave the remaining half to the great doctor and told him that even this small amount, half the size of a walnut, was more than enough. He left Helvetius with instructions, and agreed to return the next day. He never came back, and Helvetius never saw the man again. He began to doubt Elias' claims, but, at the urging of his wife, heated some lead to which he added part of Elias' stone. To their surprise, the lead became golden. They took the material to a goldsmith, who pronounced it the finest gold he had ever seen.

What is so interesting about this incident is that Helvetius was not a man known for exaggeration or flights of fancy. Indeed, as a man of medicine, he was famed for his rational approach. Perhaps the mysterious Elias, sensing the turn of the tide, approached Helvetius as if to warn him that the Art and all it stood for was in danger of being lost as science grew and threatened to erode the sacred bond between human beings and nature.

The Story of James Price

Helvetius' reputation was hardly tarnished by his admission that he believed he had achieved transmutation. The same, however, cannot be said for the English alchemist, scientist and Royal Society Fellow James Price.

The Royal Society at the time Newton took over its Presidency in 1703 was still an organization that tried to make insects from cheese, whose experiments were famously lampooned by Swift in *Gulliver's Travels*. Newton stamped out that side of the Society forever, and by the time James Price was elected a member in 1778, any mention of alchemy would invite professional and personal ruin, as Price was to discover.

Price was born in London in 1752, and excelled at Oxford, where he proved to be a potentially great chemist. He was elected into the Royal Society at the age of twenty-nine, and began a series of experiments at his house in Guildford. He was so confident that he had found the way to turn base metals into gold, that he performed a demonstration in front of a number of dignitaries. Specimens of Price's gold found their way to the King, George III, and news of his experiment began to excite the scientific community. At length, Sir Joseph Banks, the President of the Royal Society, insisted that Price repeat his experiment for the members of the Society, an offer that Price could not refuse.

Under considerable pressure, and convinced that his initial experiment could not be replicated, Price drank poison in front of the Royal Society's delegation in the course of his demonstration, and died in front of their very eyes in his laboratory. He was deemed to have lost his mind, and his passing seems a symbolic end to the history of alchemy in Europe

3: A Brief History of Alchemy in the East

The origins of Western alchemy are, as we have seen, shrouded in mystery. The beginnings of Eastern alchemy are even more obscure. The art was known to have been practiced in India, Tibet, China, Japan, Burma, Korea, Indonesia and Vietnam. Eastern alchemy is generally more concerned with the search for elixirs of longevity and vitality than attempts to transmute metals, and incorporates disciplines such as yoga, and diet and breath control.

Chinese Alchemy

It seems that Chinese alchemy is the oldest of the Eastern branches, and may even predate Western alchemy. The Chinese word for elixir, *chin-je*, may be the root of the Arabic *kimia*. The Yellow Emperor Huang Ti (2704-2595 BC) is the legendary first alchemist in China, who is said to have learnt the art from three immortal women, who also generously saw fit to instruct him in the arts of love. Chinese alchemy inherits a number of facets from folk belief, such as the idea of a plant that, when eaten, will grant immortality, together with the mystic quest for spiritual illumination.

Like their Western counterparts, the Chinese alchemists believed that they were speeding up natural processes. Work upon metals was seen as a way of freeing them from normal, linear time. The alchemical work could therefore be seen as a way to break free from time and its concomitant ravages - old age, illness, and death. The idea of the link between the microcosm and the macrocosm also appears in Chinese alchemy. In fact, the Chinese are much more specific, in that they link various organs to the elements: the heart is both fire and cinnabar, while the kidneys are equated with water and lead.

Ko Hung (254-334 AD) believed that alchemy could not be learnt from books, but required study under a master. He also advised solitude for the work, echoing a similar sentiment found in Albertus Magnus.

Taoism

The earliest Chinese alchemists seem to have been the Taoists. Tradition ascribes the origin of Taoism to Lao Tzu, a sage who lived around 500 BC, although the wisdom contained in the teachings themselves are probably much older.

Taoism, which means The Way or Path, holds that the Tao creates all things in the universe. The Tao itself is not a thing, but is to be found everywhere. (Interestingly, Paracelsus reveals himself to be familiar with this concept when speaking of the *prima materia*: he believes that all things are created from one single matter, and that this matter, the *mysterium magnum*, is the Mother of All Things.) The Tao is infinitely adaptable and sustains the world. Taoism uses the complementary principles of yin and yang to describe the properties of things: yin signifies the passive, female element, while yang is the active, male element. This idea reappears in Western alchemy under various names: sulphur and mercury, the fixed and the volatile, Sol and Luna, King and Queen. Metals were thought to be a product of *chi*, which means breath or life-force: *chi* incubates and nurtures the metals in the earth like a placenta, until they reach maturity. The Greek concept of *pneuma* is a Western equivalent, while in India, the concept of *prakrti* fulfils a similar role.

The earliest mention of alchemy in Chinese history comes from 144 BC, when the manufacture of gold was forbidden in an imperial edict. Anyone found to be manufacturing gold would be executed. However, only eleven years later, an alchemist was received at the court of Emperor Wu. The alchemist claimed to know the secret of immortality, and Emperor Wu was naturally interested. The alchemist told him to worship the Goddess of the Stove, who, when she appeared, would enable him to transmute cinnabar into gold. This gold could then be eaten, which would then lead to immortality. This story of eating gold is typical for China of the period. Commercial gain was frowned upon, while the quest for long life was seen as innately noble.

In 60 BC, the Emperor Suan granted the alchemist Liu Hsiang a license to make gold. Liu had the funds of the imperial treasury at his disposal, and proceeded to empty it in his quest for immortality. His total failure led him to be accused of breaching the edict of 144 BC, and he was sentenced to death. His sentence was only quashed when Liu Hsiang's brother paid a generous ransom of (presumably non-alchemical) gold.

The first major work on the Art is the *Ts'an T'ung Ch'I*, or *Treatise on the Three Principles* by Wei Po Yang (142 AD), which is written in the form of a commentary on the *I Ching* or *Book of Changes*. Wei writes that he lives quietly in a remote valley, having refused an offer to come to court in 121 AD. A famous story sees Wei and three disciples in the mountains. Wei decides to try his elixir on a dog, to see if it is safe for humans to take. The dog dies instantly, and the disciples are somewhat perturbed. Wei announces that to return from the mountains without the elixir would be as bad as dying from taking it, and takes the elixir himself. Needless to say, he dies too. One of the disciples is sufficiently impressed to take the elixir himself, believing that Wei had a master's reasons for taking something fatal. And he dies as well. The remaining two disciples decide to retire to the lowlands, leaving all thought of longevity behind them. As soon as they leave, Wei revives, and also resurrects the dog and the faithful disciple. He writes a letter to the other two disciples, which is delivered to them by a woodcutter they meet in the mountains. The two disciples realise that Wei is a genuine master, and are filled with sorrow at not having believed in his elixir.

Laboratory Alchemy and Spiritual Alchemy

Despite the quest for immortality through the development of elixirs and medicines in laboratory alchemy, known as *Wai Tan* alchemy, there also emerged a purely spiritual form of alchemy, known as *Nei Tan*. This is noted in the writings of Hui-ssu (515-577 AD), although it seems to have reached its apogee much later, around the ninth and tenth centuries. By the time Zen Buddhism (or Chan Buddhism, to give it its Chinese name) emerges in the thirteenth century, *Nei Tan* had become the dominant form of alchemy in China.

Like its later Western counterpart, *Nei Tan* used laboratory metaphors to describe the work. Metals were identified with parts of the body: blood and semen were seen as internal equivalents to mercury, with breath and bodily strength being lead, the matter to be transformed. *Nei Tan*'s aims were also vitality and longevity, but involved the use of yogic and meditational practices. The Chinese seem to have the edge over their Western counterparts in recognising that there was an inner dimension to the work that dealt with transforming the alchemist themselves, perhaps the true goal of the work.

Nei Tan alchemy works on three areas: *chi, Ching* and *sheen. Ching* means vital essence, and *sheen* is breath control, often linked with meditation and visualisation. Through meditation, yoga, breathing exercises and a strict dietary regimen, the Chinese *Nei Tan* alchemists aimed to keep the energy flowing through their bodies, which they saw as the key to prolonging life.

Indian Alchemy

Indian alchemy is very much influenced by Hinduism, and, like its Chinese counterpart, is concerned with the production of elixirs to prolong life. The idea of twin polarities governing things is borrowed from Hinduism: Shakti, the feminine principle, is the active, Mother/ Destroyer, the endless change of the world, whereas Shiva is the constant, passive, male energy. Its origins are contemporaneous with Chinese alchemy, and the two traditions seem to have periodically borrowed from one another.

Indian alchemy also has a number of parallels with yoga and tantra: all three ways aim to cleanse the body and mind, aiming for the so-called 'glorified body' that is impervious to time and decay. They aim for an indefinite prolongation of youth, strength and suppleness. Breath control and work with the chakras is common to both tantra and alchemy, both aiming to release hidden energies that are latent within the body in order to gain enlightenment and absolute liberty. This is reminiscent of the Western idea that the philosopher's stone is possessed by all, yet never used because the majority of people do not recognise it. The flow of energy throughout the body was thought to directly affect consciousness, so, to the Indian alchemist, to control the breath was to gain control over the way one perceived the world.

As with alchemy in China, the underlying aim of the work is to free the alchemist from the constraints of time. While the materials are being worked on, the alchemist is still in the state of becoming. Mercury has to be fixed or 'killed' (this is identified with the ego). The end of the work, in which the metals and Nature (both nature in general and the alchemist's own nature) are 'perfected', sees the alchemist reach the state of pure being, in which there is no past or future, no further striving upwards, but an abiding state of clear, mindful existence.

Amongst the oldest Indian alchemical texts are those attributed to Nagarjuna, the prominent Buddhist thinker. He is said to have attained a great age and is credited with helping a number of emerging Buddhist communities through magical means. He held that, in order to embark upon the alchemical path, one had to be devoted to the work, intelligent, without sin and in charge of one's passions. The *Rasaratnasamuccaya*, an early proto-chemical text, in a passage strikingly similar to Ko Hung and Albertus Magnus, urges that the alchemist be:

> those who are truthful, free from temptation, love the Gods and are self-controlled and used to live upon proper diet and regimen - such are to be engaged in performing chemical operations.

Echoing Albertus' dictum that the alchemist 'should reside in an isolated house in an isolated position', the *Rasaratnasamuccaya* advises that the laboratory should be set up in a forest far from all contaminating influences (other people, presumably). The alchemist must bear respect for his or her master, and must also respect the goddess Shiva, who is said to have taught the Art to mortals.

Shiva is important in Indian alchemy, as the matter the alchemist works upon is said to be *prakrti*, or primordial matter. This matter, of which all things are ultimately made, is the physical manifestation of the Goddess herself. To undertake the Great Work was therefore to intervene in the processes of Nature and Feminine forces, and to become one with them. This could be likened to the chemical wedding of the king and queen in Western alchemy (who are frequently portrayed as engaging in intercourse). The Tantric tradition would interpret this as the ultimate spiritual union of opposites through the physical union of lovers, the union in which the mercury (or ego) is finally killed, enabling the alchemist to achieve the highest freedom.

Indian alchemists also made scientific discoveries, often well in advance of their European counterparts. The importance of the colour of flame in the analysis of metals was known as early as the twelfth century. Certain metallurgical processes were described by Hindu alchemists three centuries before the work of Paracelsus, Agrippa and Agricola, and the internal use of metals was practiced a full six centuries before Paracelsus. (The modern vitamin pill with its mineral additives is a descendant of this tradition.)

The Tamil Siddhars

One of the most interesting alchemical movements in India was that of the South Indian Tamil Siddhars. They claimed descent from a sunken continent that lay in the southern part of the Indian Ocean (recent research by the Ocean Drilling Programme has revealed that the Kerguelen islands were once part of a much greater land-mass, possibly the lost continent of the Siddhars), and were masters of medicine, literature and yoga. In their texts they ridiculed society's conventions, and, like their western counterparts, seem to have operated on its fringes. The Siddhars were masters of yogic practices, and stressed that the goal of the Great Work was self-development, not the production of precious metals. As with the later Gnostic schools in the Middle East (which influenced Hellenistic alchemy), they emphasised knowing reality directly through reading the book of Nature.

According to tradition, the first Siddhar is Agastyar, a legendary sage in the same mould as Hermes Trismegistus. He is credited with teaching the arts and sciences to humanity, and is said to have been the first human to speak and write in Tamil, having been taught the language by the god Chandraswami. He is said to still live in the Pothigai Hills, from where he occasionally emerges to bestow blessings on the devout seeker.

The seventeenth century alchemist Bhogar is credited with numerous fantastic achievements, including the spreading of Siddhar wisdom into China, by managing to reincarnate as Lao Tzu. He must have been an accomplished master of reincarnation, as Lao Tzu lived some two thousand years earlier. (Perhaps even more impressively, he is said to have travelled to China by a form of primitive aeroplane.) On a more historically verifiable level, Bhogar is known to have been a prolific poet, composing what has come to be known as the '7000', a collection of seven thousand spontaneously composed verses describing his enlightenment.

Poetry: The Alchemy Of Language

Bhogar, like alchemical and mystical poets in the West, used words to both reveal and conceal; he was not necessarily using them in any accepted sense. He used them not for their logic or sense, but for their ability to hint and suggest. What for most people would simply be a means of communication, was, for the alchemical and mystical poet, a means of accessing a world infinitely more vast and replete with manifold meaning. Words are the talismans, or visible representatives, of something that is being brought forth to enrich the reader in some vital way. Poetry, the alchemists knew, could, like music, unite both heart and mind; in effect, it is the alchemical process in miniature.

Agastyar was said to have understood the secret of language, in that he knew that a form and its name are somehow insolubly linked (a link denied by Western philosophy). Alchemical language is therefore a vital component of the work as much as the laboratory operations and physical disciplines. The language of an alchemical poem literally creates the world, or conditions, necessary for the work to take place. The alchemist, therefore, lives in a world of poetic possibility in which the ability of things to transform is only limited by his or her imagination and vocabulary. Language, the alchemists knew, is one of the ways to the Stone, and one of the keys to freedom.

4: Modern Alchemy

After the death of James Price, alchemy was seemingly a lost cause. What practitioners there were in the Art kept a low profile, and as far as history was concerned, alchemy had been killed by chemistry, the rise of the sciences and the increasing dominance of the Cartesian world-view. Alchemical adepts wandered from city to city, appearing in the history books long enough to make transmutations, and then disappearing again before the inevitable arrests, kidnappings, tortures and death. A man known as Lascaris was said to have performed a transmutation for the Landgrave of Hesse in the early part of the eighteenth century, and his pupil Johann Frederick Bottger accidentally discovered porcelain in his quest for the Stone. Another story relates how the Countess of d'Erbach gave refuge to a man at her castle in Odenworld one night, and who then changed all her silver into gold as a parting gift the next morning. She never learned his true identity and she never saw him again.

However, the nineteenth century saw a revival of occult traditions, perhaps as a response to the increasing industrialization of Europe. The occult revival was not just concerned with alchemy, however, and figures such as Mrs Besant and Madame Blavatsky of the Theosophical Society and the Hermetic Order of the Golden Dawn concerned themselves with all of the hermetic arts.

In 1850, *A Suggestive Enquiry into the Hermetic Mystery* by Mary Anne Atwood was published and almost immediately withdrawn for fear that the book had given away too much. Mrs Atwood lived in Hampshire and studied the Art with her father, at whose behest the *Suggestive Enquiry* was withdrawn. (His jealousy of his daughter's achievement may have also played a part in the book's withdrawal.) Mrs Atwood regards the alchemical work as an entirely inner discipline, without the slightest hint of laboratory work. In this respect, she is following on from such great non-laboratory alchemists such as Jakob Boehme and Thomas Vaughan. She hints that the prime matter to be worked on is, in fact, the human imagination, which implies, intriguingly, that we already possess the Philosopher's Stone, the Elixir, the Great Medicine, yet are unaware of it. This would concur with the famous passage in the seventeenth century tract *The Glory of the World* that holds that *the Stone is known by all, touched by all everyday, yet is unknown.*

Mrs Atwood, in stressing the entirely inner aspect of the Work, paves the way for the purely psychological interpretation of alchemy that was first espoused by Hubert Silberer in 1914, whose work was later eclipsed by that of Carl Jung.

Jung and the Psychological Interpretation of Alchemy

Jung's interest in alchemy began around the same time that Silberer was doing his research. He kept having a dream in which he saw that his house had another wing that he had never noticed before. Jung eventually managed to gain access to this undiscovered part of the house, to find that it contained a magnificent library. Upon closer inspection, he found that the books, all of which were leather bound folios from the fifteenth and sixteenth centuries, contained alchemical diagrams and texts.

Jung began to study alchemical books during his waking hours, and believed that the alchemists had accidentally made psychological discoveries, and that the nigredo, or initial black, chaotic stage of the Work, was, in fact, the unconscious. The various stages of the Work are, according to Jung, stages in what he called individuation, or the psychological process that marks the growth of a personality into a balanced maturity.

In the first stage, the matter is cooked in the vessel. This corresponds, in Jung's view, to a personal crisis that threatens to destroy the personality. Yet, in order for the sufferer to recover fully, the personality will have to be destroyed anyway, but voluntarily. This surrender of the ego is vital to the process' ability to heal, and brings to mind the dictum stressing that, in order for the Work to be successful, the alchemist needs to be humble (ego-free).

In the later stages of the work, the self is purified, which would correspond to the albedo, or whitening, of the matter, and in the Citrinitas stage, the individual would continue on their path of recovery through learning to become grounded again. The final stage, that of Rubedo, would involve a complete integration and acceptance of the person's experience and personality. Jung held that we all go through this process many times throughout the course of our lives.

Modern Laboratory Alchemy

Although Jung's view of alchemy has prevailed in modern times - indeed, he has virtually resurrected it as a topic for serious enquiry - he has been criticised for denying the practical, laboratory aspect of the Art.

The most enigmatic and controversial modern alchemist is Fulcanelli, who was thought to have been the only practitioner of the Art to succeed in finding the Philosopher's Stone in the twentieth century. Fulcanelli's literary fame rests on two books, *The Mystery of the Cathedrals* (1926) and *The Dwellings of the Philosophers* (1930). In the first book, Fulcanelli explains how gothic cathedrals, in particular Notre Dame in Paris, were deliberately constructed to conceal esoteric secrets. He notes that a number of bas-reliefs on the cathedral show various stages of the alchemical Work. The second book continues this idea, widening the scope to study such buildings as Westminster Abbey and Holyrood Palace in Edinburgh.

To this day, his actual identity remains a mystery. The story goes that one day Fulcanelli handed his disciple, Eugene Canseliet, the manuscript of *The Mystery of the Cathedrals*, asked him to publish it, and promptly disappeared. In his introduction, Canseliet writes that 'Fulcanelli is no more', which could either mean his master was dead, or had assumed a new identity. If this were the case, it recalls the ancient Japanese practice that dictated that once one had achieved mastery in one discipline, one should abandon it, and go off to learn something new in a new town, studying under a new name.

Fulcanelli may be merely the pseudonym of a group of Parisian occultists who called themselves the Brotherhood of Heliopolis, of which Canseliet was a member. The group also included the painter Jean-Julien Champagne (1877-1932), and the scholars Rene Schwaller de Lubicz (1887-1961) and Pierre Dujols (1872-1926). Cases have been made for these three men as the authors of the Fulcanelli texts, but no one has ever succeeded in proving the case one way or the other. Fulcanelli is perhaps best left as a mystery. If he was merely a fiction created by the Brotherhood of Heliopolis, then they would have no doubt wished to create a master with fabulous powers, and, in publishing pseudonymously (the name means 'little Vulcan') would have been simply been conforming to alchemical tradition.

If, however, Fulcanelli was a real individual, then the story becomes more interesting. He was supposedly at least 70 when he gave Canseliet the book to publish, and yet, when Canseliet met Fulcanelli again in the 1950s, he had not aged. In fact, he had grown younger, appearing to be a man of about fifty years of age. Furthermore, Canseliet claims that once Fulcanelli appeared to him as a woman, the living image of Lady Alchymia. As a final, tantalising hint of the 'real' Fulcanelli, the modern adept Mark Hedsel tells of meeting a master in Florence in May 1978. During the course of the conversation, the old man asked Hedsel if he had ever met Fulcanelli. Hedsel was surprised that he should have been asked such a question, and replied that he was under the impression that Fulcanelli had died when he was still a young man. 'Oh no', the old man replied, 'Fulcanelli is alive. Fulcanelli is even older than me, yet he is still alive. He lives here in Florence.'

Alchemy and Paranoia

One of the most celebrated 'sightings' of Fulcanelli was made by the French writer Jacques Bergier. He claims that in June 1937, he met an alchemist at a chemical plant outside Paris. The stranger, a man of mature years, spoke of the dangers of atomic research, and left Bergier with the unmistakable feeling that the man, whom he took to be Fulcanelli, knew of the secrets of nuclear power eight years before the bombs were dropped on Hiroshima and Nagasaki. (Interestingly, during the development of the atom bomb, members of the Manhattan Project consulted alchemical texts, further hinting that alchemists may have known of the secrets of nuclear power for hundreds of years.) After the war, a number of international intelligence agencies, including the CIA, tried to find Fulcanelli, but without success.

There has been a long history of the relationship between alchemy and government. From the Chinese Imperial edict of 144 BC, practitioners of the Art and the rulers have been uneasy bedfellows. Usually, kings or emperors would merely request that the alchemist make gold, and force them to suffer the consequences if they failed. (Usually, this was the case, although as recently as 1706, alchemical gold was made for Charles XII of Sweden by an alchemist named Paykull: the purity of the gold was confirmed by one General Hamilton of the Royal Artillery, who was acting as an independent observer. Despite his success, Charles had Paykull executed anyway.)

If gold was not the source of official interest, then usually alchemists who were responsible for discoveries or inventions were prevailed upon to offer their services. Sometimes, these discoveries were of such magnitude that they were virtually completely suppressed, such as the rumoured invention of a primitive form of photography, first by the Arabs, and then later in the West by Leonardo da Vinci. Anything that might have had a military use was usually seized. The Dutch alchemist Cornelius Drebbel, who invented the first submarine, had to launch the vessel in the presence of King James I, who witnessed its successful maiden voyage down the Thames in 1621. Drebbel went on to work as an explosives expert for the Royal Navy, where he invented a form of torpedo, and also worked on a perpetual motion machine. The great Yugoslavian inventor Nikola Tesla (1856-1943) also worked in similar areas, and believed that he had discovered a limitless source of free energy. When he died, the FBI confiscated a large number of his papers. These have never been seen again. In 1993, rumours began to circulate that the Russian mafia was offering for sale a substance known as 'red mercury', which, according to a tip-off from an anonymous Russian scientist, was the vital ingredient in a new type of nuclear weapon, one whose production was alleged to be alchemical in nature. More recently, it has become known that the CIA for many years used 'remote viewers' during intelligence operations. The method, which recalls Roger Bacon's supposed ability to see distant events with the aid of his 'prospective glass', required a person to sit in a darkened room who then tried to 'see' what the agent, who was usually many miles, or even continents, distant, would encounter.

The story of one of the most elusive of English alchemists, John Kellerman, perfectly encapsulates the relationship between the Art and the powers that be, a relationship of unease, paranoia, secrecy and official denial. Kellerman was visited by Sir Richard Phillips, who described their encounter in his book *A Personal Tour Through the United Kingdom* (1828). Kellerman's father was said to have been a very tall Prussian, who, in order to escape being conscripted into Frederick the Great's regiment of giants, fled to the West Indies, where he married. Kellerman himself, according to Sir Richard, was well over six feet in height, and lived in isolation in the village of Lilley in Hertfordshire, a reclusive Prospero. Phillips describes Kellerman's house as being extremely untidy, littered with the usual array of retorts and crucibles, with Kellerman himself living entirely in one room. All the other rooms in the house were kept padlocked, and all the windows were boarded up. Kellerman explained that he had made gold, and that he had offered to pay off the national debt; an offer that Lord Liverpool had declined on behalf of the King. He then tried to interest the French ambassador in his discovery, who also declined. He claimed that every court in Europe knew of his secret, and that numerous attempts had been made on his life, hence the boarded up windows and padlocks. Kellerman was so concerned that the secret would fall into the wrong hands that he burnt everything he had ever written concerning the Work, and carried loaded pistols with him at all times. Of the secret, all he would say to Sir Richard was, 'The world, sir, is in my hands!' Sometime after Sir Richard's visit, Kellerman disappeared. It was rumoured that he had gone to Paris, although whether he had gone voluntarily or had been taken by force remains unclear.

Alchemy Today

Interest in the Art today is stronger now than at perhaps anytime since its mediaeval heyday. Both laboratory and philosophical, inner, alchemy have their various apologists. Since Fulcanelli, the practical side of the Art has been championed by the English physiotherapist Archibald Cockren, who followed the Paracelsian tradition of believing that the chief use of alchemy is in the preparation of medicines. The French alchemist Armand Barbault was of like mind, and produced a 'vegetable gold' that proved impossible to fully analyse when submitted to chemical tests in Germany and Switzerland; its medicinal properties were said to be remarkable, especially in the treatment of heart and kidney complaints, although its lengthy preparation eventually made it an unfeasible commercial proposition. The American tradition, inaugurated in the seventeenth century by George Starkey and John Junior Winthrop, has had its most notable descendant in the figure of Frater Albertus, whose workshop in Utah, the Paracelsus Research Institute, became the focus for a new generation of students interested in learning the Art. His *Alchemist's Handbook* has become a modern classic on laboratory procedure.

Jung has written about the inner side of the Art extensively, and the psychological approach has proved an extremely rich seam to mine. A visit to any bookshop will bear this out, with the proliferation of books entitled 'The alchemy of…' It would seem that, in the Middle Ages, the craze for alchemy may have largely been due to the quest for gold; today that would seem less likely. There is something in the Art that draws people to it, to study it, and to name their books after it.

We Are All Alchemists

The Work has been compared to many things. One of the most striking parallels, made by Paracelsus, is with the course of human life itself. The stages of life correspond to the stages of the Work in the laboratory: the initial, confused, black stage, represents our incomplete understanding of the world that we experience in adolescence and which continues for years into adulthood. The whitening phase compares to a process of maturity, in which youthful zeal gives way to a more balanced, measured approach to life. The final stage of the work, the rubedo, is left up to the individual. This is the opportunity to take one's knowledge a step further, to consciously marry the disparate elements of one's experience and weld them into a complementary whole, to fix the volatile and to volatise the fixed, or, as Thomas Vaughan wrote, to 'become transmuted into living philosophical stones'. Once this is achieved, then we have completed the Great Work; we have redeemed ourselves and, in doing so, redeemed the cosmos. Alchemists of all ages would no doubt agree with the old Hebrew proverb that 'he who saves one life saves the world entire.'

And what of the miracles, the transmutations? Did alchemists really turn lead into gold? These stories, and those of the supposed near-immortality of Artists such as the Flamels or Fulcanelli, are perhaps best likened to the Zen koan. In Zen, the koan, or riddle, is used by the master in order to get the student to stop thinking logically, and, in doing do, achieve enlightenment. When one understands that the sound of one hand clapping or whether the tree falling in the forest with no one to witness it makes a sound is not something to be measured or experienced logically, then we are perhaps nearer to understanding the meaning of the stories concerning the magical abilities of the alchemists.

However, in the long hours in the laboratory, the alchemist would no doubt have developed a very different state of consciousness to that normally experienced by most people (especially if chemical fumes were involved!), and perhaps this state, this devotion to the Work, would lead to discoveries within the alchemist, an inadvertent unleashing of some of the mind's hidden powers. Maybe these powers transmuted the lead into gold, or convinced the alchemist that they had. An incident from the early 1960s perfectly illustrates the extraordinary abilities of the power of belief over matter. An Italian by the name of Vittorio Michelli was diagnosed as having a cancerous tumour on his left hip. His condition was so bad that his hip socket had disintegrated completely, and he was sent home to die. Michelli, however, believed that if he went to bathe in the waters at Lourdes, he would be cured. Two months after returning, he was up and walking again. X-rays taken in 1965 showed that his hip-socket had completely regenerated itself. He had performed the impossible: through his faith, he had destroyed the cancer and grown new bone.

Whatever treasures genuine alchemists were said to possess, they were never said to hoard any gold that they may have produced or acquired. It is possible that they no longer needed it, or that it was merely a by-product of some inner transmutation, a profound realisation about the true nature of the world and their place in it. Fulcanelli is not alone in stressing that the alchemist must become gold themselves before any other matter can be transformed.

Lama Govinda tells a story that illustrates something similar: once in Tibet, a robber wanted to possess a legendary magic sword in order to get one over on his rivals and take over their criminal empires. A guru explained to him that, in order to acquire the sword, the robber would have to perform a very strict set of exercises: meditation, prayers, yoga, observances. So great was the robber's desire for the sword, that he agreed to perform the rituals as specified. Eventually, after years of diligent practice, the robber went to the spot where the guru had told him that the sword would appear. After the correct number of prostrations, the sword appeared before the robber. He reached out to grab the handle, and at that point realised he no longer needed it: he had achieved enlightenment, and all earthly treasures, be they magic swords or gold, were of little worth compared to the inner treasures that his practice had revealed to him.

This treasure is everywhere about us, as the great adepts never failed to point out, being handled everyday by rich and poor alike, yet hardly anyone realises its true worth. Alchemy seems to specify that we must make our way individually, working the coal of everyday life in order to find the diamond within it. If the student experiences any success, then any material treasures would be shared out, as did Nicholas and Perenelle Flamel in their gifts to church and hospital, or the Swedish alchemist Urban Hjarne, who went out into the winter streets of Stockholm to distribute medicines to the poor.

We shall leave the last word to Chaucer, whose *Canon's Yeoman's Tale* fully describes the difficulties of the path, and, indeed, of the difficulties of writing about something which was only ever passed on orally:

> Plato had a disciple once, and he
>
> Said to his master - if you care to see,
>
> It is recorded in the *Chimica*
>
> *Senioris Zadith Tabula* -
>
> 'Tell me the name, sir, of the Secret Stone?'....
>
> 'No, no', said Plato then, 'on no account.
>
> Philosophers are under strict control
>
> Never to tell that secret to a soul
>
> Or write it in a book; it is unpriced,
>
> Being a secret very dear to Christ.
>
> It is His will that no discovery
>
> Be made of it, save where His Deity
>
> Wills to inspire His servants, else forbidden.
>
> No more; from whom He wills He keeps it hidden.'
>
> (Penguin Classics version, Nevill Coghill Translation)

5: Some Alchemists

As alchemy is traditionally seen as a solitary path that does not readily lend itself to being taught in the environment of mystery schools or secret societies, it may be worth looking at some of the notable figures in the tradition. When alchemists speak of the microcosm reflecting the macrocosm, they do not just mean the relationship of the individual to nature, but also to all previous practitioners of the Art, who were said to watch over subsequent generations like guardian angels. Here are some of the more colourful names:

Abraham Eleazar (Dates uncertain; before late C14)

Nothing is known of this shady figure, other than he was reputed to be the author of the book that Nicholas Flamel purchased, *The Book of Abraham the Jew*, which inspired Flamel's alchemical quest. Further titles ascribed to Abraham are *The Book of Hieroglyphics* and the *Ancient Chemical Work*.

Georgius Agricola (1494-1555)

Agricola was a mining and mineral expert best known for his *De Re Metallica* (On Metallurgy) (1556), which was the chief textbook on mines and mining technology for many years. He is widely regarded as the father of modern Metallurgy.

Henry Cornelius Agrippa (1486-1535)

Agrippa is best known for his massive *De Occulta Philosophia* (1533) (published in English as *Three Books of Occult Philosophy*), perhaps the single most influential book of Renaissance magic. Agrippa studied with the legendary Trithemius of Sponheim, who saw the first version of *De Occulta* in 1510 and advised Agrippa against publication. It was only after Agrippa published the red herring of *De Vanitate Scientarius* in 1530, in which he attacked the occult sciences in an attempt to get church authorities off his trail, did he venture the first volume of *De Occulta* into print. Agrippa was also a respected humanist scholar, and may have been involved in espionage. He was long believed to have set up a network of secret societies across Europe, and held views about women and sexuality that were far ahead of their time. Once, when he worked for the town council of Metz, he risked his life by successfully defending a woman accused of witchcraft. Such charges could easily have been levelled against Agrippa himself, who possibly acted as one of the inspirations for Christopher Marlowe's Doctor Faustus.

Frater Albertus (1911-1984)

Pseudonym of Dr Albert Richard Riedel, one of the leading figures in twentieth century practical alchemy, who founded the highly influential Paracelsus Research Society workshop in Salt Lake City. His *Alchemist's Handbook* (first published in 1960 and reprinted many times since) has become required reading for those wishing to undertake lab work.

Albertus Magnus (1193-1280)

The great *Doctor Universalis,* who was one of the first Western adepts. His scientific work included the discovery of potassium lye. See chapter 2.

Alhazen (Ibn al-Haytam) (c. 965-1039)

Arab doctor whose book on optics *Kitab al-manazir* contains the earliest extant description of a camera obscura. The Arabs discovered photosensitive substances such as silver nitrate and silver chloride that are still used in photography today. It is possible that experiments in combining the camera obscura with these chemicals were made, which would make the Arabs the inventors of photography, believed by some to be one of alchemy's longest and best-kept secrets.

Johann Valentin Andreae (1586-1654)

Andreae was a protestant theologian who was believed to be the author of the two Rosicrucian manifestos, the *Fama* and the *Confessio*, published in 1614-15, and *The Chemical Wedding of Christian Rosenkreutz* (1616), which uses specifically alchemical symbolism in its story of the humble Christian Rosenkreutz's spiritual awakening.

Francis Anthony (1550-1623)

Anthony was an English doctor who used alchemy as a way of developing medicines. He was a follower of Paracelsus, and was repeatedly persecuted by the Royal College of Physicians. At one point he was even imprisoned, only to be freed through the intervention of his wife. Anthony was famous for his 'potable gold', an elixir that made him rich.

St Thomas Aquinas (c. 1225-1274)

Thomas Aquinas is arguably the most important figure in medieval philosophy, and one of the most influential figures in the whole of western thought. His achievement was to show that the faculties of reason were not incompatible with the Christian faith, and to play a large part in bringing the works of Aristotle into the Christian fold. His two great works, *Summa Contra Gentiles* (c.1260) and *Summa Theologica* (1267-73) contain his mature thought. In addition to these mammoth works, Aquinas wrote voluminously on Christian doctrine, Aristotle and the Arab doctors, in addition to writing polemics to order for the Church. Although today, Aquinas is regarded as doctrinally sound by the Catholic Church, in 1277, portions of his work were condemned in what was known as the Paris Condemnations (which also attacked other prominent thinkers, including Roger Bacon.) In 1323, Aquinas was declared a saint. The Condemnations were revoked two years later. For his connection with alchemy, see Chapter 2.

Arnold of Villanova (1235-1311)

Arnold was a reforming physician and alchemist born near Valencia. He was a prolific writer, although, as with many of the great names in alchemy, some of the works attributed to him are spurious. He travelled widely and is perhaps the most important exponent of medical alchemy before Paracelsus. He is known to have treated Popes Clement V and Boniface VIII, although he was briefly imprisoned in Paris for his unorthodox (i.e. strongly magical and cabbalistic) views on the Trinity. He is said to have possessed an elixir that aided rejuvenation and longevity.

Elias Ashmole (1617-1692)

Although best known for founding the Ashmolean museum in Oxford, Ashmole was also a practicing alchemist and his principle contribution to alchemy was his editing the *Theatrum Chemicum Britannicum* (1652), intended as a British 'reply' to the *Theatrum Chemicum* of 1602. Both works are essentially anthologies of previously printed alchemical works, with Ashmole collecting all the significant English works up to that point, including authors such as Thomas Norton, Thomas Charnock and Sir George Ripley. Ashmole was also the Historian of the Order of the Garter, and was a member of the 'Invisible College', which later became the Royal Society.

Mary Anne Atwood (1817-1910)

Author of *A Suggestive Enquiry into the Hermetic Mystery*, who believed that the alchemical work was entirely an inner discipline working with the imagination. Her story is fictionalised in Lindsay Clarke's novel *The Chymical Wedding* (1989).

Avicenna (Ibn Sina) (980-1037)

Known as the 'Prince of Physicians', Avicenna was an Arab alchemist and doctor. He refuted the possibility of transmutation, believing that alchemical gold was merely a clever imitation of the real thing. His views on medicine were regarded as gospel until the advent of Paracelsus.

William Backhouse (1593-1653)

A mysterious figure, chiefly known as Elias Ashmole's adoptive father. Only when he was dying in 1653 did he confided in Ashmole 'the true matter of the Philosopher's Stone', an event which Ashmole regarded as more significant than the fact that he was about to lose his father.

Roger Bacon (c. 1214-c. 1292)

After his death, the work of Roger Bacon lapsed into obscurity, no doubt due to his reputation as a sorcerer. Stories of his magical prowess kept his name alive, and in 1589, he featured in a play by Robert Greene, *The History of Friar Bacon and Friar Bungay*, which competed for audiences with Marlowe's *Doctor Faustus*. Like Faustus, Bacon regrets his magic, but unlike Marlowe's protagonist, he recants and is allowed to live (it is his slapstick servant, Miles, who is transported to hell). In a more recent tribute, Umberto Eco makes Bacon William of Baskerville's hero in *The Name of the Rose*.

Bacon remains one of the most remarkable minds of the Middle Ages, an almost clairvoyantly sighted figure centuries ahead of his time. After reading some of Roger's work, the Francis Bacon scholar Robert

Ellis was said to remark, 'I am inclined to think that he may have been a greater man than our Francis.' For more on Roger Bacon, see Chapter 2.

Armand Barbault (1900-1980)

French alchemist and astrologer whose 'vegetable gold' was found to have remarkable curative effects when subjected to laboratory tests. Its high cost and lengthy production techniques prevented it from being marketed commercially.

J.C. Barchusen (1666-1723)

Barchusen was one of the last chemists to exhibit an interest in alchemy, and worked from a laboratory set up in the city walls of Utrecht. His book *Elementa Chemiae* is full of alchemical illustrations.

Bartholomew the Englishman (fl. 1230s)

Bartholomew was one of the first great encyclopaedists, whose work *De Proprietatibus Rerum* (On the Properties of Thing) was an attempt to assimilate the new learning that was coming out of Spain as a result of the work of translators such as Robert of Chester. He appears to have studied under Robert Grosseteste, the Bishop of Lincoln, and may therefore have known Roger Bacon.

Bernard of Trevisa (fl. 1380s)

Bernard was a German alchemist who came from Trier, then known as Treves or Trevissa. He is known to have corresponded with Christine De Pisan's father, court astrologer Thomas of Bologna, who served Charles V and VI of France. (Thomas had prescribed an alchemical concoction to Charles VI and the Duke of Burgundy that had not proved too successful, and Thomas was suspected of deliberately trying to poison them. Bernard reassured Charles that the medicine was not poisonous, and Thomas escaped without punishment.) Bernard spent years on numerous experiments, spending vast sums to test the ideas of Arab alchemists such as Rhazes and Jabir; he lost so much money that he came to hold them in very low esteem. At one point he was using up to 2000 hens eggs in experiments; in another, vitriol fumes rendered him unconscious for fourteen months. At the age of 62, Bernard journeyed to Rhodes, where he encountered a certain man of religion, who agreed to fund further experiments. Bernard is said to have been finally successful in the Work at the age of 80.

Jakob Boehme (1575-1624)

Boehme was a remarkable mystic who began as a follower of Paracelsus. He worked as a shoemaker, and used alchemical ideas in his poetry and writings. With his work, alchemy becomes an entirely inner discipline.

Hermann Boerhaave (1668-1738)

Boerhaave was a Dutch chemist who made a study of alchemy, concluding that it was about as comprehensible as a work by Heraclitus.

Bolos of Mendes (fl. c.300-250 BC)

The earliest Western alchemist whose name has come down to us. For more on Bolos, see Chapter 2

Petrus Bonus (fl. 1330s)

Petrus was the author of a popular manuscript, *The New Pearl of Great Price*, which began to circulate in the 1330s. The work is notable for its honesty: Petrus admits that he has been unable to make the Stone, but believes that it is possible.

Johann Frederick Bottger (1682-1719)

German alchemist, who, after an apparently successful transmutation in the autumn of 1701, found himself on the run from both Frederick I of Prussia and Augustus II of Poland, both of whom wanted his secret. Bottger was kidnapped by Augustus' agents and imprisoned in Dresden, where, after seven years' work, he produced not the Philosopher's Stone, but the first European porcelain.

Robert Boyle (1627-1691)

One of the founders of modern chemistry, Boyle was also a practicing alchemist. See Chapter 2.

Tycho Brahe (1546-1601)

Although best known as an astronomer, Brahe was also a follower of Paracelsus, and a number of medicines that he concocted found their way into the official Danish pharmacopeia. For years Tycho also kept a detailed weather diary, convinced that weather patterns held vital secrets. It was not until 1960 that Tycho's hunch was proved right - Edward Lorenz's studies of weather patterns led to the invention of a new science: chaos theory.

Giordano Bruno (1548-1600)

Bruno was educated by the Dominicans, but had to leave the order after accusations of heresy were levelled against him. He was often in trouble, and as a result, had to travel widely. He was in England during the 1580s, where he espoused a return to Egyptian religion, and may also have been a spy for Sir Francis Walsingham, reporting on the activities of Catholic factions in England and Scotland from the vantage point of the French embassy in London. In 1592 he returned to Italy, where the Inquisition immediately arrested him. He was imprisoned and tortured for eight years, before being burnt at the stake for heresy in February 1600.

Tommasso de Campanella (1568-1639)

Author of the influential alchemical utopia, *The City of the Sun* (1602), whose admirers may have included Francis Bacon. He spent many years imprisoned by the Inquisition (he was also a supporter of Galileo), and in 1628 was prevailed upon to perform magical protection ceremonies for Pope Urban VIII, whose imminent death was being predicted by various astrologers. Urban lived for another sixteen years, and was so grateful he granted Campanella's immediate release, whereupon he fled to France.

Eugene Canseliet (1899-1982)

French alchemist. Fulcanelli's disciple, he was perhaps the only man ever to know his mysterious master's true identity. He went to his grave without ever disclosing the secret.

Jean-Julien Champagne (1877-1932)

French painter and occultist. Long suspected of being Fulcanelli, although this has never been proven. Pierre Dujols (1872-1926) and Rene Schwaller de Lubicz (1887-1961) are the other oft-mentioned names in the Fulcanelli mystery.

King Charles II of England (1660-1685)

Charles is known to have installed a laboratory beneath his bedroom. It is not known whether Nell Gwynne had access.

Thomas Charnock (c. 1524-1581)

Elizabethan alchemist who belongs to the tradition of English alchemical poetry begun by Sir George Ripley and Thomas Norton. See Chapter 2.

Geoffrey Chaucer (c. 1340-1400)

Alchemy appears in the *Canterbury Tales* in the Canon's Yeoman's Tale, in which the yeoman relates the story of his master's futile quest for transmutation. Given the detail of the story, it seems probable that Chaucer himself had studied alchemy, and may have even been a practitioner. His friend John Gower may have also been involved in the art, and may have even instructed him. (Alchemy also appears in Gower's great poem *Confessio Amantis*).

Archibald Cockren (fl. 1930s, 1940s)

English physiotherapist who, like Paracelsus before him, believed that alchemy had largely medical benefits. He described his discoveries in his book *Alchemy Rediscovered and Restored* (1940). His laboratory in Holborn was destroyed in an air raid during the Second World War. It is not known whether Cockren survived the blast. Some traditions hold that he moved to Brighton, where he is said to have died in 1950, on the verge of discovering the Stone.

Jan Amos Comenius (1592-1670)

Lutheran pastor who composed *The Labyrinth of the World and the Paradise of the Heart* during the early years of the Thirty Years' War. He eventually fled to England, where he may have participated in Samuel Hartlib's circle.

Lady Anne Conway (1642-1684)

Lady Anne was a Rosicrucian, active in Hartlib's circle. She presided over the most important esoteric centre of the day, at Ragley Hall in Warwickshire.

The Cosmopolite (fl. 1600s)

See Alexander Sethon and Michael Sendivogius.

Marie (1867-1934) and Pierre (1859-1906) Curie

Before their Nobel Prize-winning work on radium, Marie and Pierre studied alchemy. After Pierre's death, Marie was approached by a mysterious cabal who advised her to desist in her alchemical studies, which was not thought to be a subject befitting a modern French woman.

John Dastyn (fl. 1320s)

When Pope John XXII issued his Bull of 1317 condemning alchemy, it was John Dastyn (or Dasteyn) who wrote to him defending the art. Dastyn was one of the leading exponents of the art of the period, and his letter to the Pope seems to have met with some success, as when the Pope died, he left a fortune reputed to be of alchemical origin.

Arthur Dee (1579-1651)

Son of John Dee, Arthur was one of the leading English Paracelsians of the early seventeenth century. Like Francis Anthony and others, his views lead him to be persecuted by the Royal College of Physicians. He served as Royal Physician to James I, and later was summoned to Russia to serve as Tsar Michael's personal doctor.

John Dee (1527-1608)

One of the pre-eminent figures of Elizabethan England, Dee's reputation became tarnished through his involvement with the notorious fraudster Edward Kelley (1555-1595). During the 1580s they travelled in Europe (possibly involved in espionage, or the rumoured network of secret societies said to have been founded by Agrippa), and the two men conducted a series of séances through which they claimed to be in touch with angels. The angelic communications were written in an angelic language, Enochian. The séances came to an end when one of the angels suggested wife swapping. Dee was also said to have found the Philosopher's Stone in Glastonbury Abbey.

Sir Kenelm Digby (1603-1665)

Something of a dilettante alchemist, Sir Kenelm apparently turned to the Art after the death of his wife. He was involved with Hartlib's Invisible College, and at one stage embarked on a life of piracy in the Mediterranean to put himself back in funds.

Cornelius Drebbel (1572-1633)

Inventor of the first submarine, whose maiden voyage down the Thames in 1621 was witnessed by King James I. Drebbel became something of an expert on explosives, and spent most of the 1620s working for the Royal Navy, where he seems to have designed a form of torpedo. He was also known to have worked on a perpetual motion machine, and is said to have also invented the thermometer and introduced the telescope and microscope to England. His interest in the theatre (see Robert Fludd) led to his devising of a machine to that produced storms on stage, and a magic lantern for projecting images. His patron, Lord Buckingham, was assassinated after the failure of the 1627 expedition to La Rochelle, and Drebbel subsequently fell from favour. He ended up running a pub in London.

St Dunstan (909-988)

St Dunstan was the Abbot of Glastonbury Abbey from 963 until his death. He is reputed to have carried out alchemical operations in the Abbey kitchens, one of only two Abbey buildings to survive the dissolution in 1539.

Nicholas Flamel (c. 1330-1417?)

The legendary French alchemist, who was supposed to have lived to at least the age of 400. He was seen in India around the year 1700. For more on Flamel and his *soror mystica*, Perenelle, see chapter 2.

Robert Fludd (1574-1637)

English doctor whose Paracelsian leanings delayed his admittance to the Royal College of Physicians for a number of years. He was sympathetic to the Rosicrucians, and published a number of tracts in their defence. He was also passionately interested in architecture, especially in the design of theatres; the microcosm of the stage being seen as reflecting the macrocosm of the world outside the auditorium. In drama, he felt, the whole mystery of creation could be played out. He may have been responsible for designing the Globe Theatre, in which many of Shakespeare's plays were first performed.

Fulcanelli (fl. 1920s)

Pseudonym of a legendary twentieth century adept, either a genuine individual Master, or a name used by a group of Parisian occultists. (The name, meaning 'Little Vulcan', is a reference to the Roman smith-god Vulcan, one of the mythical early alchemists.) One of the more bizarre rumours about Fulcanelli suggests that he is, in fact, Nicholas Flamel, which would mean that *The Mystery of the Cathedrals* would have been published when its author was 596. For more on Fulcanelli, see chapter 4.

The Comte de St Germain (d. 1784?)

A mysterious figure who moved with apparent effortless ease through the courts of pre-revolutionary France. He was reported to possess the secret of transmutation, and at one point was involved in the manufacture of precious stones. He was reported to be of about fifty years of age in 1710, which would have put his age in 1784 as 124 (he himself claimed to be 4000). However, if someone was described as being as about fifty years of age, it was usually a coded reference to the fact that they were an initiate on a particular spiritual path. Interestingly, though, no one ever saw The Comte eat, which leads one to suspect he may have been practicing ancient Indian yogic techniques, involving strict dietary and breath control. He has been seen several times since

1784, including sightings by Madame Blavatsky in India in the late nineteenth century, and in 1973 a man claiming to be the Comte appeared on French television.

Robert Grosseteste (c. 1175-1253)

Bishop of Lincoln who was one of the leading intellectual figures in thirteenth century England. He taught Roger Bacon.

Samuel Hartlib (c. 1600-1662)

An exile from Germany courtesy of the Thirty Years' War, Hartlib was the prime mover behind the Invisible College, which attracted all the leading figures of the day, including Boyle, Ashmole and Digby. Hartlib lived just long enough to see the Invisible College receive the seal of royal approval, becoming the Royal Society two years before his death.

Jan Battista van Helmont (1579-1644)

One of the founders of modern chemistry, Helmont was the first to recognise gases and the word was, in fact, coined by him. He was nearly accused of heresy by the Jesuits after one of his papers was published in Paris in 1621 without his permission. Helmont also provides us with several accounts of transmutation: he claims that the Stone is yellow 'such as is saffron', and that it is remarkably effective in the curing of migraines and muscular ailments.

John Frederick Helvetius (1625-1709)

Helvetius also notes that the Stone is the colour of saffron; his account of transmutation is one of the most fascinating in the entire alchemical corpus. See Chapter 2.

Hermes Trismegistus

The legendary founder of western Alchemy. The writings ascribed to him were later dated by Isaac Causabon in 1614 to the first centuries after Christ, although the wisdom they contain may be much older.

Jabir (c. 721-c. 815)

Along with Roger Bacon, Paracelsus, Leonardo, van Helmont and Newton, one of the great early scientific minds. He was known as Geber in Latin, most of whose works were written by divers hands in the twelfth and thirteenth centuries. See Chapter 2

Athanasius Kircher (1602-1680)

Jesuit scholar whose wide-ranging interests, aside from alchemy, included ancient history, geology and the origins of language. Unlike his contemporary Campanella, he managed to remain unmolested by the Inquisition, never claiming to be anything other than a devout Catholic.

Abraham Lambspring (fl. c.1590s)

Nothing is known of the author of the *Book of Lambspring*, one of the most celebrated illustrated works in the alchemical corpus, other than that he may have been German. The name, a pseudonym, draws attention to the lambing season, in other words, the astrological sign of Aries, traditionally said to be the most auspicious time to begin the Great Work.

Leonardo da Vinci (1452-1519)

Although not known specifically as an alchemist, Leonardo was interested in optics, and, according to legend, produced the world's first extant photograph, the Turin Shroud, using a camera obscura and fixing the image using a concoction involving egg whites. The forgery was reputedly done on behalf of the powerful Savoy family, and after his death, a number of his notebooks - possibly those that detail his photographic experiments - were purchased by the Savoys and have never been seen again.

Rabbi Judah Loew (d. 1609)

Rabbi Loew lived in Prague in the heyday of Rudolph II and the Gold Alley. Legend ascribes to him the creation of the Golem, a servant made from mud that is animated by saying God's name. The Golem is synonymous with the alchemical homunculus, the creation of life in the retort. This can be interpreted metaphorically, of course, as meaning the dawning of spiritual understanding. It has also been seen as a proto-Frankensteinian myth, that alchemists actually produced life in the laboratory. Mary Shelley based the character of Victor Frankenstein in part on the seventeenth century Bohemian alchemist Konrad Dippel (1672-1734), who, it is claimed, created an homunculus in his castle laboratory. In more recent times, the greys of UFO lore have been interpreted as surviving golems or homunculi run amok.

Raymond Lully (c. 1235-1316)

Majorcan scholar and alchemist. One of the founding fathers of Spanish literature, he was unlikely to have written most of the numerous alchemical tracts attributed to him.

Michael Maier (1568-1622)

A German doctor and diplomat, Maier was also closely involved with the Rosicrucians. He travelled widely, and seems to have been active in implementing alchemical ideas in the political and social sphere. He authored a number of influential emblem books, the best known of which is *Atalanta Fugiens* of 1618.

The seventeenth century was the high water mark of alchemical book publishing, and, aside from Maier, other notable emblem books include Heinrich Khunrath's *The Amphitheatre of Eternal Wisdom* (1609), Johann Mylius' *Philosophia Reformata* (1622), and Daniel Stolz's rare *Viridarium Chymicum* (1624).

Moses Maimonides (1135-1204)

One of the earliest Jewish medieval scholars to write on alchemy, he may have even practiced the Art himself. His best known work is the *Guide for the Perplexed.*

Maria Prophetissa (first century BC?)

Along with the roughly contemporaneous Kleopatra, one of the mythical founding Mothers of alchemy. Although women are only occasionally prominent in alchemical records, the esteem in which Maria and Kleopatra are held testifies to the fact that alchemy respected the position and contribution of women almost from the beginning of its recorded history. See Chapter 2.

Claudio Monteverdi (1567-1643)

Monteverdi's interest in alchemy and the hermetic arts was typical of his time: all the first operas were alchemically themed, as was much drama and poetry. Monteverdi himself was a lifelong practitioner of the Art. See Chapter 2.

Morienus (fl. late 600s)

Christian hermit and alchemist who was said to have taught Prince Khalid ibn Yazid, the first Muslim alchemist. His *Book of the Composition of Alchemy* was the first alchemical tract to appear in the West, translated into Latin in 1144 by Robert of Chester.

Sir Isaac Newton (1642-1727)

Newton apparently began his alchemical experiments in late 1669, just prior to becoming Lucasian Professor of Mathematics at Cambridge (the chair now held by Stephen Hawking). He worked incessantly and reclusively until he became an MP in 1689, whereafter his punishing schedule seems to have abated. He is known to have burnt some of his papers shortly before his death in early 1727: no one has ever established what the material contained, although they may well have been of an alchemical and magical nature. Many of Newton's writings on alchemy do, however, survive, although only a fraction has been published. For more on Newton, see Chapter 2.

August Nordenskiold (1754-1792)

Finnish alchemist who studied chemistry and mineralogy in the then capital, Turku. He became a devout Swedenborgian, and, in a vision, saw the Philosopher's Stone as being able to put an end to poverty and misery, heralding a New Jerusalem (a theme famously employed by another Swedenborgian of the same time whom he may have even met, William Blake). He worked incessantly on producing the Stone, his Scandinavian origins being betrayed by his use of the metaphor of the sauna in describing the Great Work. He eventually decamped to Africa, where the conditions seemed more favourable for the founding of a Utopia (Europe then being ravaged by war and revolution). He died in Freetown, Sierra Leone.

Thomas Norton (c.1433-c.1514)

English alchemical poet. See Chapter 2.

Ostanes (fl. c.300 BC)

Persian magus and reputedly Alexander the Great's personal alchemist. He features as a character in Bolos of Mendes' *On Natural and Initiatory Things*.

Paracelsus (c. 1493-1541)

The so-called 'Christ of Medicine', a major medical reformer and iconoclast, whose many enemies put it about that his death in 1541 was due to a drunken orgy (he may have in fact either been assassinated or died of rickets). The word 'bombastic' derives from Paracelsus: it was his nature, as well as being one of his middle names. A huge school sprang up after his death (see below). For more on his work, see Chapter 2.

Paracelsians (fl. 1560s-1600s)

Paracelsus' influence grew vastly after his death, and his followers exerted a profound influence on European medicine in the remaining years of the sixteenth century and well into the seventeenth. Among his most prominent followers were Adam von Bodenstein (1528-1577), who edited many of the great man's works and published them at Basle, Gerard Dorn (fl. 1560s-1580s), Joseph Duchesne (1521-1609), Petrus Severinus (1542-1602), the Rulands, both called Martin (Senior died in 1602, Junior in 1611), and the polemicist Andreas Libavius (1560-1616), who managed to inadvertently further the cause of the Paracelsians through his vigorous attacks on them.

In England, the cause was supported by men like Francis Anthony, Robert Fludd and John Dee's son Arthur. The Royal Society of Physicians made a point of persecuting anyone suspected of Paracelsian sympathies (both Anthony and Fludd clashed with the Society). Other names in this field include John French (fl.1640s), Thomas Thymme (d.1620), who knew John Dee and translated Duchesne into English, and Nicholas Culpepper (1616-1654), the celebrated herbalist.

Eirenaeus Philalethes (fl. 1660s)

See George Starkey

Eugenius Philalethes (fl. 1650s)

See Thomas Vaughan

Giambattista della Porta (1535-1615)

Della Porta was one of the leading lights of seventeenth century natural magic, whose book *Magiae Naturalis* (1558) dealt with magic, spiritualism and scientific experiments in addition to alchemy. He is also known to have been the first to produce tin monoxide. He 'perfected' the camera obscura, a device known to the Arabs as early as the eleventh century; after its first public demonstration della Porta was thought to have produced the images by magic and was arrested on a charge of sorcery.

Francois Rabelais (c. 1494-1553)

Rabelais' great satire, *Gargantua and Pantagruel*, published in five volumes between 1532 and 1564, espouses humanistic learning and the famous vision of Utopia in the Abbey of Theleme sequence. Although seen as a literary buffoon - frequently portrayed as drunk, he cuts an image similar to that of his great contemporary, Paracelsus - Rabelais was almost certainly involved with the hermetic arts. His writing betrays a mastery of the Language of the Birds, and he himself states that he is writing in code, urging his readers to sniff the books like a dog with a bone, until they 'break the bone and suck the substantific marrow - that is to say, the meaning of the Pythagorean symbols that I employ'.

Sir Walter Ralegh (c. 1554-1618)

When Ralegh was awaiting execution in the Tower of London (the site of Raymond Lully's supposed transmutations for Edward II), he was apparently persuaded to perform alchemical experiments by one Mrs Hutchinson, the wife of the superintendent of the Tower. She wanted to keep the great man occupied during his incarceration, and also wanted to put his great knowledge to practical use: she is said to have distributed the medicines made by Ralegh in his experiments to the poor.

Rhazes (866-925)

Arab doctor who lectured at Ray and at the hospital in Baghdad. 21 alchemical works are ascribed to him, including the *Secret of Secrets*, which was known to Roger Bacon.

Sir George Ripley (d. 1490)

English adept who is said to have studied on Rhodes with the Knights of St John. He is responsible for transmitting the secret (perhaps indirectly) to both Thomas Norton and Thomas Charnock. See Chapter 2.

Tiphaigne de la Roche (fl. 1720s)

French alchemist whose book *Giphantie* details his experiments in photography.

Emperor Rudolph II (1576-1612)

Holy Roman Emperor who held court in Prague and was famously sympathetic to alchemists. He sheltered a number of people from church persecution, including John Dee, and was hostile to Rome for most of his reign.

Michael Scot (c. 1175-c. 1235)

Scottish scholar and translator of Aristotle. He served the Holy Roman Emperor Frederick II as court astrologer, and has the distinction of appearing as a character in both Dante's *Inferno* and Boccaccio's *Decameron*.

Alexander Sethon (d. 1603/4) and Michael Sendivogius (1566-1646)

Sethon emerges into history briefly in 1601, when he rescued the crew of a Dutch ship foundering off the coast of Scotland. He visited the captain of the ship in Holland the following year, and performed a transmutation for him. Sethon then became something of an alchemical missionary, and travelled through Europe demonstrating his powers to anyone who took an interest. He was imprisoned by Christian II, Elector of Saxony, and was tortured. He was rescued from jail by Sendivogius, but only lived another year or so after his ordeal. He refused to divulge the secret to Christian, but apparently left some of his elixir to Sendivogius, who also seems to have married Sethon's widow. Sendivogius has been painted as something of a charlatan, publishing Sethon's works as his own and also appropriating his pseudonym, the Cosmopolite. However, he is said to have discovered oxygen, a discovery that was later put to use on Drebbel's submarine.

Mary Sidney (1561-1621)

Patron of the arts, translator, scholar and second Countess of Pembroke. According to John Aubrey, Mary, Sir Philip Sidney's sister, was also alleged to have practiced the Art, and Philip's dedication of *The Countess of Pembroke's Arcadia* to her was entirely fitting: the work is full of hermetic allusions. She was instrumental in having her brother's poem published after his death in 1586.

George Starkey (1627-1665)

Starkey was born in Bermuda and educated at Harvard. Along with Winthrop, who may have been his Master, he was one of the earliest alchemical adepts in the New World. He came to England in 1650, bringing with him tracts ascribed to one Eirenaeus Philalethes, whom he claimed gave him the manuscripts before his departure. The works, such as *An Open Entrance to the Closed Palace of the King* and *Ripley Reviv'd*, were to become highly influential, with alchemists holding him in high regard right down to the present day. Philalethes' identity has never been established: Starkey himself is the most likely author of the texts, although his contention that they were written by a New England doctor by the name of Childe is difficult to disprove entirely. He was a skilled chemist and was a member of Hartlib's Invisible College. He died after performing an autopsy on a plague victim.

Suidas (fl. ninth century)

Monk and encyclopaedist, who is one of the first western writers to mention alchemy.

Pope Sylvester II (Gerbert d'Aurillac c.940-1003)

Sylvester II, the first French pope, was known to have studied various hermetic arts in Spain before his Papacy. This would make him, along with Abbot Dunstan of Glastonbury, one of the earliest alchemists in the west. Despite his reputation as a necromancer (he was said to have made a pact with the devil to assure his rapid promotion to the Pontificate) he was also a strong Pope who was a tireless supporter of science, music and mathematics, and was a pioneer of the abacus, terrestrial and celestial globes and preserving the manuscripts of classical antiquity.

Solomon Trismosin (fl. c.1480-1500)

Pseudonym of the author of *Splendour Solis*, one of the most beautiful of all alchemical illustrated books, the earliest surviving copy of which dates from 1532. Trismosin is also held by some to have instructed Paracelsus.

Trithemius (1462-1516)

One of the most mysterious figures in Renaissance magic, Trithemius was the abbot of the monastery of Sponheim in Germany. A pioneer of modern cryptography, his codes were mistaken for satanic invocations, and he may have been one of the inspirations for Marlowe's Faustus. Luckily, Sponheim was remote enough to ensure that the Inquisition didn't come for him, and he avoided the stake. He is said to have taught Agrippa and Paracelsus.

Basil Valentine (fl. 1394-1413)

A number of influential books bearing Valentine's name appeared in Germany in the closing years of the sixteenth century. He is said to have been the prior of Erfurt at the end of the fourteenth century. In keeping with alchemical tradition, his name is a pseudonym: according to Fulcanelli, the name derives from Greek word *basileus* (king) and the Latin *valens* (powerful). There has long been speculation about whether he really existed, or whether his work was actually written by Johann Tholde, who edited the first printed editions. Tholde was a salt-boiler who had strong Rosicrucian connections and may have been a practicing alchemist himself.

Anthony Van Dyck (1599-1641)

Painter and alchemist. He was, in all probability, introduced to the Art in 1623 by a man who later became his patron, Sir Kenelm Digby.

Thomas Vaughan (1621-1665)

English alchemical poet who wrote under the name of Eugenius Philalethes. His twin brother was the metaphysical poet Henry Vaughan. He is said to have studied the Art with his wife Rebecca, who died in 1658.

Wei Po Yang (fl. 120-140)

Chinese adept, author of the oldest extant alchemical work in China, the *Ts'an T'ung Ch'i*. See chapter 3.

John Junior Winthrop (1606-1676)

Winthrop has the claim to being the first in several fields: he was the first alchemical adept in America; he was the first Governor of Connecticut; he was the Royal Society's first colonial member, becoming elected in 1663; and was the first astronomer of note in the New World. He was also a successful doctor (he was a follower of Paracelsus), and taught George Starkey.

Denis Zachaire (fl. mid-sixteenth century)

Denis left an autobiography detailing his lengthy quest after the Stone. This was first printed in 1583, and was reprinted in a number of languages, attesting to its popularity. He reveals the problems he had in securing finance, avoiding the superstitions of his neighbours, the hostility and dismay of his family, who saw him squandering all the money intended for his education in a single year of experiments, the sudden death of patrons and masters (one of whom seems to have died from inhaling soot in the laboratory, whose atmosphere, he records, was akin to 'the arsenal of Venice, where cannon are cast'), to say nothing of frequent outbreaks of plague and an ultimately successful attempt on Denis' life (this chapter is known to have been written by Mardochee de Delle, court poet to Rudolph II). It remains uncertain whether this was a real account, or a satire based on the eventful life of the long suffering and (one hopes) eventually successful Artist, Bernard of Trevisa.

Zosimos of Panoplis (fl. c.300)

Hellenistic alchemist, and one of the most revered of early practitioners. He may have worked in conjunction with his sister, Theosebeia. See Chapter 2.

Suggestions For Further Reading

This list is deliberately brief, as the number of books on alchemy is vast. However, those listed below are ones that should be reasonably easy to get hold of.

General History

Michael Baigent and Richard Leigh *The Elixir and the Stone* Penguin Books 1997

Mircea Eliade *The Forge and the Crucible* (1956) University of Chicago 1978

Cherry Gilchrist *The Elements of Alchemy* Element Books 1993

Mark Haeffner *Dictionary of Alchemy* HarperCollins 1992

E.J. Holmyard *Alchemy* (1957) Dover Books 1990

F. Sherwood Taylor *The Alchemists* (1947) Barnes and Noble 1992

C.J.S. Thompson *The Lure and Romance of Alchemy* (1932) Bell Publishing 1990

Psychological

Titus Burckhardt *Alchemy: Science of the Cosmos, Science of the Soul* (1962) Fons Vitae 2000

Marie Louise von Franz *Alchemy - An Introduction to the Symbolism and the Psychology* Inner City Books 1980

Carl Gustav Jung *Psychology and Alchemy* (1944) Routledge 1990

Richard and Iona Miller *The Modern Alchemist* Phanes Press 1994

Jay Ramsay *Alchemy - The Art of Transformation* HarperCollins 1997

Illustrated

Johannes Fabricius *The Alchemists and their Royal Art* (1976) Equation 1994

Stanislas Klossowski de Rola *Alchemy: The Secret Art* Thames and Hudson 1973

Stanislas Klossowski de Rola *The Golden Game* Thames and Hudson 1988

Alexander Roob *Alchemy and Mysticism* Taschen 1995

Gareth Roberts *The Mirror of Alchemy* British Museum 1994

Writings by Alchemists

Henry Cornelius Agrippa *Three Books of Occult Philosophy* Llewellyn Publications 1993

Frater Albertus *The Alchemist's Handbook* Samuel Weiser 1995

Elias Ashmole (Ed) *Theatricum Chemicum Britannicum* Kessinger Publishing 1998

Mary Anne Atwood *A Suggestive Enquiry into the Hermetic Mystery* (1850) Yogi Publication Society

Roger Bacon *Opus Majus* Kessinger Publishing 1990

Roger Bacon *The Mirror of Alchymy* Holmes Publishing Group 1992

Roger Bacon *The Magical Letter of Roger Bacon* Holmes Publishing Group 1988

Armand Barbault *Gold of a Thousand Mornings* (1969) Neville Spearman 1975

Archibald Cockren *Alchemy Rediscovered and Restored* (1940) Kessinger Publishing 1995

Fulcanelli *The Mystery of the Cathedrals* (1926) Brotherhood of Life 2000

Fulcanelli *The Dwellings of the Philosophers* (1930) Archive Press 1999

The Alchemical Works of Geber Trans Richard Russell Samuel Weiser 1994

Abraham Lambspring *The Book of Lambspring* (1599)

Michael Maier *Atalanta Fugiens* (1618) Phanes Press (with cassette) 1989

Adam McLean *A Commentary on the Mutus Liber* (1677) Phanes Press 1991

Paracelsus *Selected Writings* (Ed. Jolande Jacobi) (1942) Princeton 1995

Paracelsus *Hermetic and Alchemical Writings* Alchemical Press 1992

Walter Scott (Trans) *Hermetica: The Writings Attributed to Hermes Trismegistus* Solos Press 1993

A.E. Waite (Ed) *The Hermetic Museum* (1893) Samuel Weiser 1999

Related Interest

Peter Ackroyd *The House of Doctor Dee* Penguin Books 1994

Geoffrey Chaucer *The Canterbury Tales* Penguin Books 1977

Lindsay Clarke *The Chymical Wedding* (1989) Picador 1990

Paolo Coelho *The Alchemist* (1988) HarperCollins 1993
Patrick Harpur *Mercurius* Macmillan 1990
Mark Hedsel *The Zelator* Century 1998
Kenneth Rayner Johnson *The Fulcanelli Phenomenon* Neville Spearman 1980
Ben Jonson *Ben Jonson's Plays and Masques* Norton 1979
Michael Talbot *The Holographic Universe* HarperCollins 1991
E.V. Westacott *Roger Bacon in Life and Legend* Banton Press 1993
Frances Yates *The Rosicrucian Enlightenment* Routledge 1993
Frances Yates *Giordano Bruno and the Hermetic Tradition* (1964) University of Chicago 1991
Marguerite Yourcenar *The Abyss* (1968) Penguin Books 1984

Internet

Many Texts, Images and Articles can be found on the alchemy web-site run by Adam McLean at http://www.levity.com/alchemy. This is an invaluable resource for those wishing to study alchemical writings and images at first hand.

The Essential Library

New This Month:

Steven Spielberg (£3.99)
Sherlock Holmes (£3.99)
Feminism (£3.99)
Alchemists & Alchemy (£3.99)

Film Directors:

Jane Campion (£2.99)
Jackie Chan (£2.99)
David Cronenberg (£3.99)
Alfred Hitchcock (£3.99)
Stanley Kubrick (£2.99)
David Lynch (£3.99)
Sam Peckinpah (£2.99)
Orson Welles (£2.99)
Woody Allen (£3.99)
John Carpenter (£3.99)
Joel & Ethan Coen (£3.99)
Terry Gilliam (£2.99)
Krzysztof Kieslowski (£2.99)
Sergio Leone (£3.99)
Brian De Palma (£2.99)
Ridley Scott (£3.99)
Billy Wilder (£3.99)

Film Genres:

Film Noir (£3.99)
Horror Films (£3.99)
Spaghetti Westerns (£3.99)
Heroic Bloodshed (£2.99)
Slasher Movies(£3.99)
Vampire Films (£2.99)

Miscellaneous Film Subjects:

Steve McQueen (£2.99)
The Oscars® (£3.99)
Marilyn Monroe (£3.99)
Filming On A Microbudget (£3.99)

TV:

Doctor Who (£3.99)

Literature:

Cyberpunk (£3.99)
Hitchhiker's Guide (£3.99)
Terry Pratchett (£3.99)
Philip K Dick (£3.99)
Noir Fiction (£2.99)

Ideas:

Conspiracy Theories (£3.99)
Nietzsche (£3.99)

Available at all good bookstores, or send a cheque to: **Pocket Essentials (Dept AA), 18 Coleswood Rd, Harpenden, Herts, AL5 1EQ, UK**. Please make cheques payable to 'Oldcastle Books.' Add 50p postage & packing for each book in the UK and £1 elsewhere.

US customers can send $6.95 plus $1.95 postage & packing for each book to: **Trafalgar Square Publishing, PO Box 257, Howe Hill Road, North Pomfret, Vermont 05053, USA**. e-mail: tsquare@sover.net

Customers worldwide can order online at **www.pocketessentials.com**.